T0073665

Accelerating Expansion

Accelerating Expansion

Philosophy and Physics with a Positive Cosmological Constant

GORDON BELOT

OXFORD
UNIVERSITY PRESS

Great Clarendon Street, Oxford, OX2 6DP,
United Kingdom

Oxford University Press is a department of the University of Oxford.
It furthers the University's objective of excellence in research, scholarship,
and education by publishing worldwide. Oxford is a registered trade mark of
Oxford University Press in the UK and in certain other countries

© Gordon Belot 2023

The moral rights of the author have been asserted

All rights reserved. No part of this publication may be reproduced, stored in
a retrieval system, or transmitted, in any form or by any means, without the
prior permission in writing of Oxford University Press, or as expressly permitted
by law, by licence or under terms agreed with the appropriate reprographics
rights organization. Enquiries concerning reproduction outside the scope of the
above should be sent to the Rights Department, Oxford University Press, at the
address above

You must not circulate this work in any other form
and you must impose this same condition on any acquirer

Published in the United States of America by Oxford University Press
198 Madison Avenue, New York, NY 10016, United States of America

British Library Cataloguing in Publication Data
Data available

Library of Congress Control Number: 2023936729

ISBN 978-0-19-286646-2

DOI: 10.1093/oso/9780192866462.001.0001

Printed and bound by
CPI Group (UK) Ltd, Croydon, CR0 4YY

Statique sans l'être, vide mais non neutre, virtuellement actif sur toute matière qu'on voudrait y mettre, résultat d'une symétrie en trompe-l'oeil, solution bâtarde d'une équation bâtarde, l'Univers de de Sitter était donc un curieux complexe d'équivoques, qui cependant portait l'avenir de la pensée cosmologique.

J. Merleau-Ponty

Contents

Preface

A little more than a hundred years ago, in the paper in which he founded relativistic cosmology, Einstein proposed altering the equations of general relativity by the addition of a term involving the cosmological constant.[1] He was driven to this by his desire to construct an eternal and unchanging model of the Universe in which space is finite, unbounded, and full of matter: when $\Lambda > 0$, spacetime has a natural tendency towards expansion, used in Einstein's model to precisely counterbalance the tendency of matter to collapse under gravity. Interest in the cosmological constant soon flagged: the observed expansion ruled out a static Universe, undercutting Einstein's original motivation for introducing Λ; and big-bang cosmological models were available that allowed observation to be accommodated without recourse to a cosmological constant.

Twenty-five years ago, astronomers discovered, to their surprise, that the rate of expansion of the Universe appears to be accelerating. Within general relativity, a positive cosmological constant provides the simplest and most natural way to accommodate this behaviour. So the cosmological constant is back.

A central topic in philosophy of space and time is how our notions of time, geometry, and physics are changed as we move from Newtonian physics to special relativity to general relativity. In light of the developments just mentioned, it is also natural to ask how, if at all, our notions of time, geometry, and physics should alter when we consider general relativity with a positive cosmological constant.

It is obvious that there will be some changes. In the $\Lambda = 0$ regime, Minkowski spacetime (the setting of special relativistic physics) is at centre stage and provides the point of departure for much of our thinking about general relativity. When Λ is positive, this role is taken over by de Sitter spacetime. Minkowski spacetime and de Sitter spacetime differ in many crucial ways. To give just one example: whereas any event in Minkowski spacetime can send a signal to any given eternal free falling observer, this

---

[1] The cosmological constant is conventionally denoted either by a Greek letter lambda, either in upper-case (Λ) or in lower-case (λ) form.

is not true in de Sitter spacetime. It follows that Einstein's observer-relative notion of simultaneity must behave very differently in de Sitter spacetime than it does in Minkowski spacetime.

Further, there is a sense in which de Sitter spacetime plays a more central role when $\Lambda > 0$ than Minkowski spacetime plays when $\Lambda = 0$: it appears that typical $\Lambda > 0$ cosmological models are de Sitter-like in at least one temporal direction—whereas $\Lambda = 0$ cosmological models are not in general very Minkowski-like. So, roughly speaking at least, to ask what general relativistic worlds are like when $\Lambda > 0$ is to ask what de Sitter-like worlds are like.

No one will mistake this strange book for the Platonic form of a philosophical monograph. It is perhaps somewhere between a textbook and a shaggy attempt to present a snapshot of a field in which open problems extend as far as the eye can see. It descends from lectures given at the International Summer Institute in Philosophy of Physics on Philosophy of Quantum Gravity in Williams Bay in 2016 and at the Rotman Summer Institute in Philosophy of Cosmology in Goderich in 2018. The lectures were originally entitled "Ten Weird Things about de Sitter Spacetime(s)" (a title eventually dropped because I never could settle on a canonical ten). The lectures, which were aimed at graduate students in philosophy, aspired to make a start on the project of understanding the implications for accelerating expansion for the philosophy of space and time. So they focused on de Sitter spacetime and on the de Sitter-like spacetimes that are central to contemporary cosmology. They were structured around two main goals. On the one hand, I sought to present a quick and digestible overview of some relevant bits of mathematics and physics that can be hard to find in one place. On the other hand, I wanted to suggest some research topics for the students in the audience. So I spent some time discussing some of the more philosophically provocative ideas that circulate in the contemporary literature on cosmology in what was intended to be a non-judgemental fashion. The idea was to leave to students so-inclined the critical evaluation, defence, demolition, or elaboration of those ideas. This written version aims to retain those features of the lectures.

The cover image was painted by Kaća Bradonjić during my lecture in Williams Bay and appears here by her kind permission. The epigraph, which appears here by permission of Éditions Gallimard, is from Merleau-Ponty, *Cosmologie du XXᵉ siècle,* 61. The elegant line drawings from the first chapter of Schrödinger's *Expanding Universes* are reproduced here with permission from Cambridge University Press as Figures 1.1, 1.6, 2.1 and 2.4. They were drawn by Alfred Schulhof, then a young neighbour of Schrödinger's in Ireland.[2] Figure 6.1 is adapted from figure 3.3 of Griffiths and Podolský, *Exact Space-Times in Einstein's General Relativity* and appears here by permission of Cambridge University Press. A couple of the other figures are in the public domain (Figure 1.7, a rendering of the Poincaré disk by Parcly Taxel, and Figure 9.1, from a paper of Boltzmann). The rest I have drawn using my rudimentary grasp of the vector graphics languages Asymptote and TikZ (whose name is a recursive acronym, standing for *TikZ ist kein Zeichenprogramm*).

I would like to thank Nick Huggett and Chris Wüthrich for the invitation to lecture at their summer school in Williams Bay and to thank Chris Smeenk and Jim Weatherall for the invitation to lecture at their summer school in Goderich—both were models of what such occasions should be. I would also like to thank the members of the audiences at the summer schools for their engagement with my over-stuffed lectures. Thanks also to those who attended other presentations of this material in my graduate seminar on philosophy of cosmology at the University of Michigan, in a masterclass at the University of Bristol, and in talks at the Black Hole Initiative at Harvard and for the Caltech Philosophy of Physics Group.

Thanks also to Mike Schneider for helpful comments on an early and chaotic version of the lecture notes. I am grateful to Dave Baker, John Earman, and Laura Ruetsche, individually, for many things—and grateful to them, collectively, for clubbing a complete manuscript draft and for suggestions that led to many improvements. Many thanks, also, to two anonymous and remarkably supportive (if understandably bemused) readers for the Press who offered manifold astute and helpful suggestions that likewise lead to many improvements.

I am very grateful to Peter Momtchiloff for his interest in this unusual project and to Cathryn Steele for all of her help in seeing it through. Thanks also to Bruce and Kathy Craig, paragons of rationality and kindness, for their hospitality over the years. No thanks are due to Bob Batterman, who could have helped but didn't. Big thanks are due to Grisbi[†] and Cholmondeley, who

---

[2] You can read a bit about Schulhof in Moore, *Schrödinger*, and in Holfter and Dickel, *An Irish Sanctuary*—and also, I believe, in Hinks, "West Vancouver Nonagenarian Sets Swimming World Records."

helped by being (as was only right) clownish, stubborn, and demanding of attention.

My obsession with matters de Sitter had its first glimmerings in late 2015. There have been some dark days since then and some bright ones. I am more grateful than I can say to have spent them with Laura.

Skylonda
January 2023

Introduction

De Sitter spacetime began its career in general relativity as a counterexample. In his 1917 paper on cosmology Einstein espoused the Machian thesis that:

> In a consistent theory of relativity there can be no inertia *relatively to* "space," but only an inertia of masses *relatively to one another*.[1]

Einstein goes on to explain that in his attempts to construct a static, spatially infinite cosmological model meeting this standard he had encountered an insuperable difficulty: if the density of matter falls off at spatial infinity, then there will have to be a fixed empty-space background geometry that obtains in that limit—and that geometry will play a role in determining the inertia of bodies, so that we cannot say that the inertia of each body depends only on the locations and movements of the others. But Einstein had managed to find a static, homogeneous, matter-filled spacetime in which space had the structure of a three-sphere: the Einstein static universe. This spacetime was not a solution of the field equations he had published in November 1915. So Einstein proposed to modify those equations by adding a term depending on the cosmological constant Λ.[2] In a letter to de Sitter, Einstein explained the significance that he attached to the combination of the new field equations and the idea of a spatially finite cosmos.

> From the standpoint of astronomy, of course, I have erected but a lofty castle in the air. For me, though, it was a burning question whether the

---

[1] Einstein, "Cosmological Considerations in the General Theory of Relativity" (6.43), published on 15 February 1917 (see the initial note in the References for conventions for citing Einstein's papers and correspondence). For the role of this and related ideas in Einstein's thought in the period 1912–1918, see Belot, *Cosmological Reconsiderations*.

[2] The revised field equations are now usually written in the form:

$$R_{ab} - \frac{1}{2}Rg_{ab} + \Lambda g_{ab} = 8\pi G T_{ab},$$

where g_{ab} is the metric tensor, R_{ab} and R are its Ricci tensor and scalar curvature, G is Newton's constant, and T_{ab} is the stress-energy tensor of any matter fields.

Accelerating Expansion: Philosophy and Physics with a Positive Cosmological Constant. Gordon Belot, Oxford University Press. © Gordon Belot 2023. DOI: 10.1093/oso/9780192866462.003.0001

relativity concept can be followed through to the finish or whether it leads to contradictions. I am satisfied now that I was able to think the idea through to completion without encountering contradictions. Now I am no longer plagued with the problem, while previously it gave me no peace. Whether the model I formed for myself corresponds to reality is another question, about which we shall probably never gain information.[3]

It did not take long for de Sitter to write with some news that disturbed the peace: contrary to Einstein's expectations, it was possible to construct a vacuum solution of the new field equations—and de Sitter had done so.[4] Einstein initially disputed de Sitter's physical interpretation of his new solution, maintaining that it contained singularities that should be interpreted as material bodies. And he went on to reiterate his Machian position:

> In my opinion, it would be unsatisfactory if a world without matter were possible. Rather, the $g^{\mu\nu}$-field should *be fully determined by matter and not be able to exist without the matter*. This is the core of what I mean by the requirement of the relativity of inertia.... To me, as long as this requirement had not been fulfilled, the goal of general relativity was not yet completely achieved. This only came about with the λ term.[5]

A battle royal ensued, with Weyl joining Einstein in claiming that de Sitter's solution was singular and Klein joining de Sitter in arguing that it was not.[6] Einstein and Weyl each eventually yielded to Klein, admitting that de Sitter's solution was homogeneous and free of singularities—but going on to claim that because time behaved oddly in de Sitter's world, it could never be taken seriously as a physical possibility.[7] Einstein and Weyl were

---

[3] Letter of before 12 March 1917 (8.311). In this period Einstein frequently spoke of the cosmology paper as establishing the ultimate consistency of general relativity: see, further, the letters to Besso of 9 March 1917 (8.306), to Levi-Civita of 2 August 1917 (8.368), to Förster of 16 November 1917 (8.400), to Humm of 18 January 1918 (8.440), to Sommerfeld of 1 February 1918 (8.453), to Mie of 22 February 1918 (8.470), and to Klein of 24 March 1918 (8.492).

[4] Letter to Einstein of 20 March 1917 (8.313). See also de Sitter, "On the Relativity of Inertia."

[5] Letter to de Sitter of 24 March 1917 (8.317). See also two related papers of early 1918, "On the Foundations of the General Theory of Relativity" (7.4) and "Critical Comment on a Solution of the Gravitational Field Equations Given by Mr. De Sitter" (7.5).

[6] For the tangled history of this episode, see: the editorial headnote "The Einstein-de Sitter-Weyl-Klein Debate" in the *Collected Papers of Albert Einstein* (8, 351–357); Janssen, "'No Success Like Failure...': Einstein's Quest for General Relativity," §5; Smeenk, "Einstein's Role in the Creation of Relativistic Cosmology," §4; and Goenner, "Weyl's Contributions to Cosmology."

[7] For discussion and references, see Section 2.1 (for Einstein) and Section 7.1 (for Weyl).

wrong again: de Sitter spacetime played a prominent role in early relativistic cosmology. In 1923 Weyl himself proposed a cosmology in which half of de Sitter spacetime represented an expanding universe with Euclidean spatial slices, eternal to both the past and the future—a proposal that was later taken up by proponents of the steady state theory.[8] Lemaître investigated solutions that were asymptotic to the Einstein static universe in one temporal direction and asymptotic to de Sitter spacetime in the other.[9] In fact, throughout the 1920s, de Sitter spacetime was the basis for the most widely discussed cosmological models, before eventually being displaced by the matter-filled expanding solutions of Friedmann and Lemaître.[10]

Enthusiasm for the cosmological constant gradually waned. Einstein's own enthusiasm declined (non-monotonically) until it was finally extinguished when he recognized that observation strongly favoured an expanding Universe.[11] Over time, observation showed that Λ must be remarkably small in magnitude, if non-zero. The most influential textbooks of the final third of the twentieth century present Einstein's field equations of 1915 as canonical and make only a few side remarks about the amended equations.[12] And with the decline in interest in Λ came a correlative decline in interest in de Sitter spacetime—despite its simplicity, elegance, and historical importance it is usually mentioned only in passing in textbooks of this period.[13]

Things changed in the final years of the century. The most natural response to the 1998 observation of the anomalous acceleration of local galaxies is to take the Universe as a whole to be undergoing accelerated

[8] See Weyl, "On the General Theory of Relativity" and *Raum-Zeit-Materie* (fifth edition), §39 and Anhang III. On the rise and fall of the steady state theory, see Kragh, *Cosmology and Controversy*.

[9] Lemaître, "A Homogeneous Universe of Constant Mass and Increasing Radius Accounting for the Radial Velocity of Extra-Galactic Nebulae." Solutions of this kind also figure in Eddington, "On the Instability of Einstein's Spherical World" and are now generally known as Eddington-Lemaître solutions. For a number of years, Eddington promoted them as realistic cosmological models—see, e.g., *New Pathways in Science*, 220.

[10] For discussion of this transition, see Nussbaumer and Bieri, *Discovering the Expanding Universe*; and Realdi, "Relativistic Models and the Expanding Universe."

[11] See Einstein, "Zum kosmologischen Problem der allgemeinen Relativitätstheorie" and "On the So-Called Cosmological Problem."

[12] See Weinberg, *Gravitation and Cosmology*, 155; Hawking and Ellis, *The Large Scale Structure of Space-Time*, 73; Misner, Thorne, and Wheeler, *Gravitation*, §17.2; Sachs and Wu, *General Relativity for Mathematicians*, §6.2.1(f); and Wald, *General Relativity*, 99. For a collection of some mid-century expressions of antipathy towards the cosmological constant, see Peebles, *Cosmology's Century*, 57.

[13] See Weinberg, *Gravitation and Cosmology*, 615; Misner, Thorne, and Wheeler, *Gravitation*, 745 and 758; Sachs and Wu, *General Relativity*, ∅; and Wald, *General Relativity*, 109 and 116. Hawking and Ellis are an exception to this trend—they include a classic review of de Sitter and anti-de Sitter geometry in §5.2 of *Large Scale Structure*.

expansion.[14] A positive cosmological constant provides the most straight-forward mechanism for accelerating expansion in general relativity: in the presence of such a constant, this behaviour will be found in generic solutions of the equations of the theory.[15]

This puts de Sitter spacetime back at centre stage—for two reasons. On the one hand, de Sitter spacetime is the simplest and most natural example of a $\Lambda > 0$ solution, just as Minkowski spacetime is the simplest and most natural solution when $\Lambda = 0$. On the other hand, there is good evidence that should accelerated expansion continue, what we see will eventually become more and more like de Sitter spacetime: when $\Lambda > 0$, de Sitter geometry is a powerful dynamical attractor in general relativity, in a way that Minkowski spacetime is not in the $\Lambda = 0$ regime. For these reasons, de Sitter spacetime today provides a natural starting point for anyone interested in understanding what general relativity tells us about our world.

My goal here is to provide a primer on de Sitter spacetime and on the de Sitter-like spacetimes that dominate contemporary relativistic cosmology, aimed at readers who already have some knowledge of general relativity. I have tried to write the sort of overview that I wish I could have found when I first started to think about these issues. It can be thought of as an introduction to geometry, physics, and philosophy in the realm of a positive cosmological constant. The focus will initially be on de Sitter spacetime itself. Later, focus will shift to the range of de Sitter-like spacetimes relevant to cosmology—and our initial study of de Sitter geometry will pay dividends in helping us to understand some of the puzzles surrounding universes undergoing accelerating expansion.

A motivating idea of the present work is that just as every presentation of general relativity includes some discussion of Minkowski spacetime and advanced presentations also discuss spacetimes that asymptotically approach

---

[14] This step is optional for those willing to deny that the Universe is spatially homogeneous and isotropic at the relevant scales—see e.g., Smoller, Temple, and Vogler, "An Alternative Proposal for the Anomalous Acceleration."

[15] For an overview of the current evidence in favour of a cosmological model that is spatially flat and driven by a positive cosmological constant, see Peebles, *Cosmology's Century*, chapter 9. Note, however, that there is emerging evidence that some mechanism for accelerating expansion more complex than a simple cosmological constant may be required: observation of the motions of galaxies suggests one value for the Hubble constant, analysis of cosmic background radiation another. For a survey of approaches, see Di Valentina *et al.*, "In the Realm of the Hubble Tension." For philosophical discussion, see Smeenk, "Trouble with Hubble." It may be that de Sitter-like geometries are about to be banished from the limelight once again. Or it could be that this anomaly will evaporate or require only a small adjustment of the current cosmological paradigm—for optimism on this point, see e.g., Peebles, *Cosmology's Century*, 74.

Minkowski geometry as one proceeds to infinity along spacelike or null geodesics, now that we believe that we live in a $\Lambda > 0$ world, de Sitter spacetime and asymptotically de Sitter spacetimes ought to enjoy something like the same pride of place. This might seem overwrought. After all, there is a sense in which *every* Lorentz manifold is infinitesimally Minkowskian and locally approximately Minkowskian.[16] And, more generally, if Λ is small, then it ought to be safe to ignore it for many purposes:

> Evidently, even if $\Lambda \neq 0$, Λ is so small that it is totally unimportant on the scale of a galaxy or a star or a planet or a man or an atom.[17]

All of this is true—properly understood. But while there is a sense in which if you zoom in smaller and smaller neighbourhoods of a point in de Sitter spacetime, what you see looks more and more like tiny patches of Minkowski spacetime, there is also a sense in which there is always a gulf of fixed size separating the geometry of a patch of de Sitter spacetime, no matter how small, from the geometry of any patch of Minkowski spacetime (it may help to think of the ways in which zooming in will make parts of a sphere more similar to parts of a plane and ways in which it will not). In Minkowski spacetime, the sectional curvature vanishes at each point, while de Sitter spacetime is a space of constant sectional curvature $k > 0$. So no matter what patch you look in, you will find constant sectional curvature k: there is no sense in which the curvature properties of de Sitter spacetime approach those of Minkowski spacetime as you zoom in on smaller and smaller patches. And as for galaxies and stars: one might well expect that it should be possible to approximate by a $\Lambda = 0$ model any reasonably realistic $\Lambda > 0$ model of an astrophysical process that unfolds on a reasonably short timescale. But in practice, our analyses of collisions producing measurable gravitational waves are not realistic in the requisite sense, since they treat our observations as taking place at null infinity.[18] And there are crucial structural differences between null infinity in a $\Lambda = 0$ spacetime and null infinity in a $\Lambda > 0$

[16] At any point in any Riemannian or Lorentz manifold, we can choose coordinates in which the metric tensor looks like a Euclidean or Minkowski metric at that point—and hence, so long the metric tensor is continuous, we can make its components as close as we like to those of a Euclidean or Minkowski metric by zooming in on a suitably small neighbourhood. See, e.g., O'Neill, *Semi-Riemannian Geometry*, 72 f.
[17] Misner, Thorne, and Wheeler, *Gravitation*, 411. On the question of effects on cosmic expansion on local physics, see Carrera and Giulini, "Influence of Global Cosmological Expansion in Local Dynamics and Kinematics."
[18] For a textbook treatment, see Poisson and Will, *Gravity*.

spacetime—no matter how small we make our positive Λ.[19] So, like it or not, we ignore Λ at our peril.

COMING ATTRACTIONS. The first two chapters below introduce the hero of our story, de Sitter spacetime. Chapter 1 focuses on elementary facts about its geometry and symmetries. Like Minkowski spacetime, de Sitter spacetime is as symmetric as can be. In particular, for any two freely falling observers, there is an isometry of de Sitter spacetime that maps one to the other. So simultaneity is not absolute in de Sitter spacetime any more than it is in Minkowski spacetime. Chapter 2 is concerned with disanalogies between the nature of time in Minkowski spacetime and in de Sitter spacetime. Whereas in Minkowski spacetime the choice of a freely falling observer determines a natural notion of simultaneity, in de Sitter spacetime this is not true: time is stranger in de Sitter spacetime than in Minkowski spacetime. There are, however, geometrically natural subregions of de Sitter spacetime in which time and simultaneity are as well-behaved as one could wish—and these will play an important role in the discussion of Chapters 6 and 9.

The next three chapters form a bridge between the discussion of de Sitter spacetime in the initial chapters and the discussion of asymptotically de Sitter spacetimes in the final chapters. Chapter 3 surveys some results about symmetry and about spaces of constant curvature. The primary goal is to equip the reader with concepts and results that will play a role in later chapters (some readers will want to skip or skim this chapter, returning to it as necessary later). Another goal is to place de Sitter spacetime in context. A theme of this chapter is that de Sitter spacetime has a near relative, elliptic de Sitter spacetime, that is in several senses its rival—each has a claim to be the most natural general relativistic spacetime in the $\Lambda > 0$ regime. Chapter 4 offers an overview of the geometry of elliptic de Sitter spacetime. This chapter can be omitted without loss of continuity. But I hope that readers will find that it advances their understanding of de Sitter spacetime itself as well as providing insight into what is in many ways the most natural example of a temporally non-orientable spacetime, elliptic de Sitter spacetime. The chapter ends with a discussion of the reasons that physicists have had for taking elliptic de Sitter spacetime seriously as a physical model, despite its temporal non-orientability. Chapter 5 is devoted to the $\Lambda < 0$ analogs of de Sitter spacetimes: anti-de Sitter spacetime (the most natural solution of the $\Lambda < 0$ vacuum Einstein field equations) and certain spacetimes that share its local

---

[19] On this point, see Ashtekar, "Implications of a Positive Cosmological Constant for General Relativity."

geometry. It introduces readers to various facts about anti-de Sitter geometry that will play a role in later chapters. It also includes extended discussion of ideas that will play important roles through the following chapters, including the technique of conformal completion and the AdS/CFT correspondence (a profound conjectured correspondence between conformal field theories and the asymptotically anti-de Sitter sector of quantum gravity).

In the remaining chapters, we turn our attention to asymptotically de Sitter geometries. We are particularly interested in general relativistic worlds, such as we take our own to be, that become increasingly de Sitter-like as $t \to \infty$. Chapters 6 and 7 are primarily concerned with ways of understanding the notion of an asymptotically de Sitter geometry and with the grounds that we have for thinking that de Sitter spacetime is a powerful dynamical attractor in the $\Lambda > 0$ setting. By way of providing context, we will also have a look at some related questions about the $\Lambda < 0$ and $\Lambda = 0$ cases.

Chapter 8 is concerned with various senses in which cosmic topology may be underdetermined by observation. We pay special attention to some classic results due to Glymour and to Malament and to some new results due to Ringström. The focus is on cosmologically relevant examples in which underdetermination is a concomitant of exponential expansion.

Chapter 9 is devoted to the cosmologists' favourite skeptical worry about the reliability of evidence and the possibility of knowledge, the problem of Boltzmann brains: its origins in ideas of Boltzmann, its development and apparent resolution by Eddington, and its return to prominence in recent years, driven by physicists' attempts to make coherent sense of a $\Lambda > 0$ reality.

If there is a moral to this work it is that philosophers should think more about the geometry and physics of de Sitter spacetime and its relatives. They are weird.

EXERCISES AND QUESTIONS. Each of these chapters will include some challenges to enjoy, labelled either as *Exercises* or *Questions*. Exercises are supposed to be (more or less) manageable mathematical problems. They vary quite a bit in difficulty—I try to give some rough guidance by labelling them (with considerable arbitrariness) *easier, medium,* or *harder.*[20] The purpose of the exercises is to offer readers so-inclined a means of testing and advancing their understanding of material discussed in the main text. Questions raise philosophical problems that, to my mind at least, deserve attention. Mostly these are not meant to be straightforward. Again, there is a lot of variation

---

[20] Hints are provided for many of them in footnotes.

in how challenging (and also in how open-ended) they are meant to be—I have tried to indicate my own sense of this using (variants on) the labels *smaller* and *larger*. To me, at least, the answers are usually not obvious—and even in the cases where I have some confidence about what answer I would give, I would not expect other philosophers to agree with me. The balance shifts from exercises in early chapters towards questions in the later chapters, with Chapter 9 offering little by way of answers and ending with a flurry of questions. More generally, throughout my aim is to interest readers in the many philosophical puzzles about geometry, time, and physics that arise in the $\Lambda > 0$ domain, rather than to resolve such puzzles.

CONVENTIONS. The following are in force unless otherwise noted. Manifolds are without boundary and are connected (unless we are thinking of them as Lie groups—this should be clear from context). Lorentz signature is $(-, +, \ldots, +)$. We use the notation ∂_x for the partial derivative operator $\frac{\partial}{\partial x}$, ∂_{tt} for the partial derivative operator $\frac{\partial^2}{\partial t^2}$, and so on.

1
Our Hero

1. Introduction

It is time to properly introduce our main character: de Sitter spacetime. The following exposition is of necessity incomplete—some aspects will be treated in detail, others recounted quickly or passed over in silence.[1]

One of our themes in the first half or so of this book is going to be that de Sitter spacetime and its closest associates (elliptic de Sitter spacetime and anti-de Sitter spacetime) stand to Minkowski spacetime as the classical non-Euclidean geometries (spherical, elliptic, and hyperbolic geometry) stand to Euclidean geometry. There are, of course, intrinsic methods for characterizing these non-Euclidean geometries—e.g., treating them as metric spaces or via traditional synthetic systems of axioms. But for many purposes, it turns out to be illuminating to characterize spherical geometry and hyperbolic geometry as substructures of higher-dimensional Euclidean or Minkowski spaces.[2] In Section 2 below, we will take the corresponding approach to characterizing de Sitter spacetime (anti-de Sitter spacetime will be given the same treatment in Chapter 5). In Section 3 below we will be concerned with the symmetries of de Sitter geometry, focusing on the striking fact that it is homogeneous but not stationary. Section 4 will introduce some basic facts about the conformal completion of de Sitter spacetime, which will play a large role in our discussion of asymptotically de Sitter spacetimes in chapters 6–9 below.

---

[1] The classic introduction to de Sitter geometry is Schrödinger, *Expanding Universes*, chapter I. Other helpful overviews: Hawking and Ellis, *The Large Scale Structure of Space-Time*, §5.2; Griffiths and Podolský, *Exact Space-Times in Einstein's General Relativity*, chapter 4; Callahan, *The Geometry of Spacetime*, §5.3; and Moschella, "The de Sitter and Anti-de Sitter Sightseeing Tour."

[2] For this approach—and discussion of its roots in nineteenth-century geometry—see Ratcliffe, *Foundations of Hyperbolic Manifolds*, chapters 1–3.

Accelerating Expansion: Philosophy and Physics with a Positive Cosmological Constant. Gordon Belot, Oxford University Press. © Gordon Belot 2023. DOI: 10.1093/oso/9780192866462.003.0002

2. De Sitter : Minkowski :: Sphere : Plane

2.1 The Plane and the Sphere

For any $d \geq 2$, the most basic example of a Riemannian manifold of dimension d is d-dimensional Euclidean space, $\mathbb{E}_d$. Each $\mathbb{E}_d$ is of course flat (i.e., has vanishing sectional curvature).[3] And each $\mathbb{E}_d$ is as symmetric as can be: the group of symmetries of d-dimensional Euclidean space has dimension $d(d+1)/2$, the maximum possible for any Riemannian or Lorentz manifold.[4]

Amongst the most basic examples of spaces with non-trivial curvature are the d-dimensional spheres. We can think of the sphere $S_d(r)$ of dimension d and radius $r > 0$ as the subset of $\mathbb{E}_{d+1}$ consisting of points lying r units of distance away from some arbitrarily chosen origin O, equipped with the metric g_S induced by the ambient Euclidean metric. According to g_S, in order to find the distance between two points in $S_d(r)$, you find the length of the shortest path in $S_d(r)$ that connects them, where the length of a path γ in $S_d(r)$ is just the length of γ considered as a curve in $\mathbb{E}_{d+1}$.[5]

For our purposes it often suffices to consider $S_d(1)$, which we will denote by S_d. On the other hand, we will use S^d to denote the d-sphere as a topological space—so S_d is S^d with the special metric just introduced (often called the *round sphere metric*).

The symmetries of $S_d(r)$ correspond to reflections and rotations of the ambient Euclidean space that fix the origin O. Like $\mathbb{E}_d$, $S_d(r)$ is maximally symmetric (the group of rotations of $(d+1)$-dimensional Euclidean space has the same dimension as the group of isometries of d-dimensional Euclidean space).

Special subsets of spheres arise by taking intersections of $S_d(r)$ with linear subspaces of $\mathbb{E}_{d+1}$ that pass through O. Geodesics in $S_d(r)$ arise as the intersection of $S_d(r)$ with the planes through the origin of $\mathbb{E}_{d+1}$ (think of the great circles of a two-sphere). Points p and q on $S_d(r)$ are *antipodal* if

---

[3] Some basic facts about sectional curvature and its relation to Riemannian curvature are reviewed in Section 3.2 below.

[4] Maximally symmetric spaces will be discussed in Section 3.3 below.

[5] Is it okay to reduce the Riemannian geometry of the sphere to a bunch of facts about distances between points? Yes. (1) Let d be a bijection between Riemannian manifolds (M, g) and (N, h) and suppose that d preserves lengths of curves (i.e., if $\gamma : [0, 1] \to M$ is a curve, then the length of γ according to g is the same as the length of $d \circ \gamma$ according to h). Then d is a Riemannian isometry—see, e.g., Petersen, *Riemannian Geometry*, §5.6.3. (2) Furthermore, Riemannian manifolds can be characterized directly as a subclass of path metric spaces (without mentioning charts, atlases, tensors, etc.)—see Gromov, *Metric Structures for Riemannian and Non-Riemannian Spaces*, 85 f.

there is a line through the origin in $\mathbb{E}_{d+1}$ on which they both lie (think of the poles of a two-sphere). Points p and q are antipodal if and only if the Euclidean distance between them is $2r$.

The two-sphere is *geodesically convex*: for any two points on the sphere, there is a geodesic on which they both lie. Indeed, if π is a plane in the ambient Euclidean space that includes p, q, and the origin O, then the intersection of π with the sphere will be a geodesic of the sphere that includes p and q—and any such geodesic must arise as the intersection of such a π with the sphere. If p and q are not antipodal (i.e., if the Euclidean distance between them is less than $2r$), then there will be a unique such π, so p and q will be joined by a unique geodesic. But for antipodal points p and $-p$, there are infinitely many such π, so pairs of antipodal points on the sphere are joined by infinitely many geodesics. This picture carries over to higher dimensions: each d-sphere ($d \geq 2$) is geodesically convex, with antipodal pairs of points being joined by infinitely many geodesics and all other pairs of points being joined by a single geodesic.

2.2 Minkowski Spacetime and de Sitter Spacetime

In any spacetime dimension $d \geq 2$, the most basic example of a Lorentz manifold is the Minkowski space $\mathbb{M}_d$ of dimension d (we will often use $\mathbb{M}$ to denote $\mathbb{M}_4$). Each $\mathbb{M}_d$ is of course flat (i.e., has vanishing sectional curvature). And each is maximally symmetric (i.e., its group of symmetries has dimension $d(d+1)/2$). Further, in each dimension, Minkowski spacetime is the only solution of the $\Lambda = 0$ vacuum Einstein equations with this degree of symmetry.

Amongst the most basic examples of Lorentz manifolds with non-trivial curvature are the d-dimensional de Sitter spacetimes. We can think of the de Sitter space $dS_d(r)$ of dimension d and radius r as the subset of $\mathbb{M}_{d+1}$ consisting of points lying r units of spacelike distance away from some arbitrarily chosen origin O, equipped with the Lorentz metric g_{dS} induced by the ambient Minkowski metric (see Figure 1.1). According to g_{dS}, in order to find the (timelike, null, or spacelike) length of a path γ in $dS_d(r)$, one calculates the (timelike, null, or spacelike) length assigned to γ by the Minkowski metric of the ambient $\mathbb{M}_{d+1}$.[6] For our purposes, it often suffices

---

[6] Is it okay to reduce the structure of a Lorentz manifold to facts about lengths of paths? Yes. (1) Let d be a map between Lorentz manifolds (M, g) and (N, h). If d is a homeomorphism that preserves lengths of timelike curves when restricted to suitably small geodesic normal

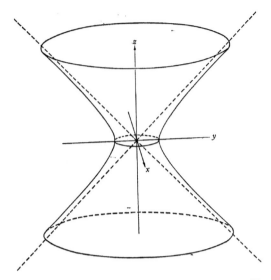

Figure 1.1 Two-dimensional de Sitter spacetime as represented by the hyperboloid $-z^2 + x^2 + y^2 = 1$ in three-dimensional Minkowski spacetime. Note that the hyperboloid is asymptotic to the past and future null cones of the origin. Of course, de Sitter spacetime does not have a beginning or ending: whenever it is pictured as a hyperboloid with a vertical axis of symmetry, one should imagine the hyperboloid being infinitely extended upwards and downwards. (This diagram is reproduced, by permission of Cambridge University Press, from Schrödinger, *Expanding Universes*, 5.)

to consider $dS_d(1)$, which we will denote dS_d—and we will often use dS unadorned to denote $dS_4(1)$. Each de Sitter spacetime $dS_d(r)$ is a globally hyperbolic solution of the vacuum Einstein equations with cosmological constant $\Lambda = 3/r^2$.

The symmetries of $dS_d(r)$ correspond to symmetries of the ambient Minkowksi spacetime that fix the origin O—i.e., all reflections, rotations,

---

neighbourhoods, then d is an isometry—see Peleska, "A Characterization for Isometries and Conformal Mappings of Pseudo-Riemannian Manifolds." And if (M, g) is strongly causal and d preserves lengths of timelike curves then d is an isometry—see Beem, "Homothetic Maps of the Space-Time Distance Function and Differentiability." (2) It would appear that, roughly speaking at least, as Riemannian manifolds stand to path metric spaces, well-behaved Lorentz manifolds stand to the Lorentzian length spaces of Kunzinger and Sämann, "Lorentzian Length Spaces." So, in parallel with point (2) of fn. 5 above, one can hope for a characterization of (well-behaved) Lorentz manifolds as special Lorentzian length spaces (without requiring the apparatus of differential geometry).

and boosts that fix O (so the symmetry group of $dS_d(r)$ can be identified with the Lorentz group of the ambient $\mathbb{M}_{d+1}$).

Special subsets of de Sitter spacetime arise by taking intersections of $dS_d(r)$ with linear subspaces of $\mathbb{M}_{d+1}$ that pass through O. Points p and q on $dS_d(r)$ are *antipodal* if there is a line through the origin in $\mathbb{M}_{d+1}$ on which they both lie. Geodesics in $dS_d(r)$ arise as the intersection of $dS_d(r)$ with the planes through the origin of $\mathbb{M}_{d+1}$, where the resulting geodesic is spacelike/null/timelike according to whether the plane in question is spacelike/null/timelike.[7] And the Cauchy surfaces of minimum volume of $dS_d(r)$ arise by taking the intersection of $dS_d(r)$ with flat d-dimensional spacelike hyperplanes of $\mathbb{M}_{d+1}$ that pass through the origin (see Figure 1.2). Any set of inertial coordinates on the ambient Minkowski spacetime determines a distinguished such Cauchy surface: the *equator* Σ_0, that arises as the intersection of the de Sitter hyperboloid with the $t = 0$ Minkowski hyperplane.

Of course, a familiar fact of life about special relativity carries over to the present context: care must be taken in interpreting pictorial representations of relativistic spacetimes, as such diagrams are inevitably misleading in

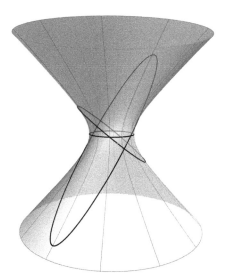

Figure 1.2 Three spacelike surfaces in de Sitter spacetime that arise as the intersection of the hyperboloid with flat spacelike hyperplanes through the origin of the ambient Minkowski spacetime.

---

[7] See, e.g., proposition 4.28 in O'Neill, *Semi-Riemannian Geometry.*

certain respects. In Figure 1.1, the equator of the hyperboloid looks like a special spatial slice. And it is: since it arises as the intersection of the hyperboloid with a flat hypersurface through the origin of the ambient Minkowski spacetime, it is a Cauchy surface of minimal volume. But from the appearance of the diagram, one might well think that this slice was the unique slice of minimal volume. But that cannot be right, since other slices also arises as the intersection of the hyperboloid with flat hyperplanes through the origin of the ambient Minkowski spacetime (see Figure 1.2). Since the de Sitter hyperboloid is invariant under Lorentz transformations of the ambient Minkowski spacetime and since such flat Minkowski hypersurfaces are related to one another by Lorentz transformations, their intersections with the hyperboloid must have the same geometry as one another.

Null geodesics of de Sitter spacetimes have an alluring feature. For ease of visualizability, consider dS_2 as a subset of $\mathbb{M}_3$ (the same line of reasoning will work for any $dS_d(r)$). Fix inertial coordinates t, x, and y on $\mathbb{M}_3$ and consider the null hyperplane π determined by the condition $x = t$. Let ℓ_1 be the set of points in π satisfying $y = 1$ and let ℓ_{-1} be the set of points in π satisfying $y = -1$. Each of ℓ_1 and ℓ_{-1} is a null geodesic of $\mathbb{M}_3$. And each also lies in the unit hyperboloid dS_2—since any point in either line will satisfy $-t^2 + x^2 + y^2 = 1$. Now, the intersection of π with dS_2 consists of two curves, each of which is a null geodesic of dS_2. So ℓ_1 and ℓ_2 are null geodesics of both $\mathbb{M}_3$ and dS_2. And since any null plane through the origin of $\mathbb{M}_3$ will be related to π by a Lorentz transformation, it follows that every null geodesic of dS_2 is also a null geodesic of the ambient Minkowski spacetime (see Figure 1.3).

De Sitter spacetimes are not geodesically convex. In a de Sitter spacetime, some pairs of distinct points cannot be connected by any geodesic, some can be connected by exactly one geodesic, and some can be connected by infinitely many. The following batch of exercises asks you to work out what is going on in some detail.[8]

EXERCISE 1.1 (Easier). Let p and $-p$ be antipodal points in dS_d.

a) Show that p and $-p$ are spacelike-related in the ambient Minkowski spacetime.[9]

---

[8] The exercises lead you along a quite elementary path. For a more sophisticated approach, see Wolf, *Spaces of Constant Curvature*, lemma 11.2.3; for helpful background, see Ratcliffe, *Foundations of Hyperbolic Manifolds*, §3.2.

[9] HINT: consider the line through the origin of the ambient Minkowski spacetime that includes both p and $-p$.

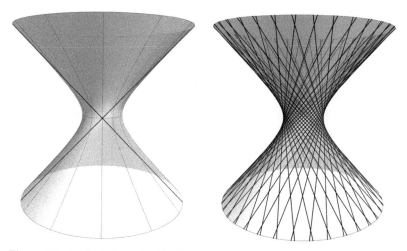

Figure 1.3 On the left: a pair of null geodesics through a point on the equator of the hyperboloid. On the right: three dozen such pairs.

b) Show that there are infinitely many spacelike de Sitter geodesics that include both p and $-p$.[10]

c) Show that there is no timelike or null de Sitter geodesic includes both p and $-p$.[11]

EXERCISE 1.2 (Easier). Let p and q be distinct, non-antipodal points in dS_d. Show that there is at most one de Sitter geodesic that includes both p and q.

EXERCISE 1.3 (Medium). Suppose that p and q are timelike related points in dS_d. Let π be the unique Minkowski plane that includes p, q, and the origin. The intersection of π with dS_d is a pair of timelike geodesics γ_1 and γ_2. Show that p and q must both lie on γ_1 or both lie on γ_2.[12]

EXERCISE 1.4 (Medium). Let p and q be null-related points in dS_d. Let π be the unique Minkowski plane that includes p, q, and the origin. The intersection of π with dS_d is a pair of null geodesics γ_1 and γ_2 that are parallel

---

[10] HINT: p and $-p$ determine a spacelike geodesic of the ambient $\mathbb{M}_{d+1}$; consider a spacelike Minkowski geodesic ℓ^* orthogonal to ℓ at the origin O.

[11] HINT: such a γ would also be a timelike or null curve in the ambient $\mathbb{M}_{d+1}$.

[12] HINT: show that no point lying on γ_1 can be timelike-related in $\mathbb{M}_{d+1}$ to a point lying on γ_2, and vice versa.

as geodesics of the ambient Minkowski spacetime. Show that p and q must both lie on γ_1 or both lie on γ_2.[13]

EXERCISE 1.5 (Medium). Suppose that p and q are non-antipodal spacelike related points in dS_d. Let π be the unique Minkowski plane that includes p, q, and the origin.

    a) Show that p and q can be chosen so that π is spacelike in the ambient Minkowski spacetime. Show that in this case p and q can be connected by a unique de Sitter geodesic, which is spacelike.

    b) A timelike (null) plane through the origin in $\mathbb{M}_{d+1}$ intersects dS_d in a pair of timelike (null) de Sitter geodesics γ_1 and γ_2. Show that no two points lying on one of these geodesics can be spacelike related.

    c) Show that p and q can be chosen so that π is null or timelike in the ambient Minkowski spacetime. Show that in this case there is no de Sitter geodesic that includes them both.

EXERCISE 1.6 (Medium). Suppose that p and q are points of dS_d that are not connected by any de Sitter geodesic and let $-q$ be the point antipodal to q. Show that some timelike or null geodesic passes through both p and $-q$.[14]

3. Homogeneous But Not Stationary

Recall that that we say that a Riemannian or Lorentz manifold (M, g) is *homogeneous* if for any two points of M, there is an isometry of (M, g) that maps one to the other. In particular, when a space or spacetime is homogeneous, its local geometry looks the same at each point.

As noted above, just as the isometry group of the sphere S_d is the group of rotations and reflections of the ambient $\mathbb{E}_{d+1}$ that fix the origin (i.e., the orthogonal group in $d+1$ dimensions), so the symmetry group of dS_d is the group of isometries of the ambient $\mathbb{M}_{d+1}$ that fix the origin (i.e., the Lorentz group in $d+1$ dimensions).

Of course, the sphere with its inherited metric is homogeneous: for any two points on S^d we can find a rotation of the ambient $\mathbb{E}^{d+1}$ that fixes the origin and maps the first point to the second point.

---

[13] HINT: show that no point lying on γ_1 can be null-related in $\mathbb{M}_{d+1}$ to a point lying on γ_2, and vice versa.

[14] HINT: $-q$ lies in the Minkowski plane π determined by p, q, and the origin.

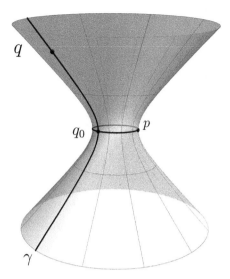

Figure 1.4 Any two points on the equator of the de Sitter hyperboloid are related by a Minkowski rotation. Any two points on a geodesic orthogonal to the equator are related by a Minkowski boost.

De Sitter spacetime is likewise homogeneous. For ease of visualization we specialize to the case of dS_2 (see Figure 1.4). Let p and q be points on the de Sitter hyperboloid. We can choose inertial coordinates t, x, and y in the ambient Minkowski spacetime so that p lies in the $t = 0$ hyperplane. So relative to these coordinates, p is in the equator Σ_0 of dS. If q is also in Σ_0, then there is a rotation of $\mathbb{M}_3$ that maps p to q (since then both lie in the same plane at the same spacelike distance from the origin). If q is not in Σ_0, then there is a timelike geodesic γ passing through q that is orthogonal to Σ_0 (let the coordinates of q be (t_1, x_1, y_1) and consider the Minkowski plane determined by the origin, q, and $(-t_1, x_1, y_1)$). Let q_0 be the point of intersection between Σ_0 and γ. Note that any two points on γ are related to one another by a boost of $\mathbb{M}_3$: to see this, note first, that performing a rotation, if necessary, we can take q_0 to have coordinates $(0, 1, 0)$ so that γ takes the form $\gamma : t \in \mathbb{R} \mapsto (t, \sqrt{1 + t^2}, 0) \in \mathbb{R}^3$; then note that any point $\gamma(t)$ is then related to q_0 by a boost in the x-direction with velocity $t/\sqrt{1 + t^2}$. So there is a Lorentz transformation that maps p to q: the result of composing a rotation mapping p to q_0 with the boost mapping q_0 to q.

Another interesting symmetry property for Lorentz manifolds is stationarity. Recall that if G is a group of isometries of a Riemannian or Lorentz manifold (M, g), then the *orbit* of G through $p \in M$ is the set of points of the form $q = g(p)$ with $g \in G$. The orbits of G form a partition of M (i.e., each point of M lies in exactly one orbit). When G is a one-parameter group of isometries (acting continuously), each orbit will either be a single point or (the image of) a curve. We say that the Lorentz manifold (M, g) is *stationary* if there is a one-parameter group G of isometries whose orbits are all timelike curves. In this case, the orbits of G correspond to a family of observers who see the (local) geometry of space as constant in time.[15]

One might naively expect that any homogeneous spacetime would also have to be stationary. For if the local *spacetime geometry* is the same at each point, how could any family of observers whose worldlines fill the spacetime think that *spatial geometry* was anything other than constant in time? But this is too quick. Choose inertial coordinates on the ambient Minkowski spacetime and consider a family of observers who are at rest relative to one another as they pass through the equator determined by these coordinates (see Figure 1.5). The observers in this family will see the local *spatiotemporal* geometry as the same at each point on their worldlines. But they will not see the *spatial* geometry as unchanging—rather they will see space as exponentially expanding towards the future and the past (since the natural notion of space for these observers will pick out the horizontal cross-sections of the de Sitter hyperboloid).

Indeed, de Sitter spacetime is not stationary: there is no family of observers whose worldlines fill the spacetime and are orbits of a one-parameter group of de Sitter isometries. To make this plausible, note that the Lorentz group of the ambient Minkowski spacetime is generated by rotations and boosts. The orbits of one-parameter groups of rotations will be spacelike (see the horizontal slices in Figure 1.5). And a boost in a given spatial direction will always leave invariant any spacelike line orthogonal to this direction (see Figure 1.6). So one-parameter groups of boosts will always have fixed points—so their orbits will not all be timelike curves.[16]

---

[15] Many authors require stationary spacetimes only to admit timelike Killing fields (or even to admit such Killing fields in a neighbourhood of spatial infinity). Our approach is close to that of Choquet-Bruhat, *General Relativity and the Einstein Equations*, §XIV.1—although she imposes further conditions to ensure that it is globally true that observers adapted to the symmetry see space as unchanging in time. In this connection, see also Sánchez, "On the Geometry of Static Spacetimes," proposition 3.3.

[16] This falls short of a proof because although the Lorentz group is generated by rotations and boosts (in the sense that its Lie algebra admits a basis consisting of the infinitesimal generators of rotations and boosts) there are the Lorentz transformations (and one-parameter groups of Lorentz transformations) that are not (composed of) rotations or boosts—on this point, see, e.g., Crampin and Pirani, *Applicable Differential Geometry*, §8.7. For a proof that no spacetime

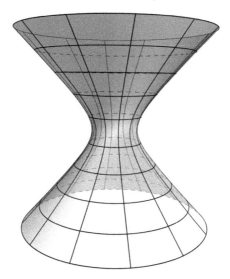

Figure 1.5 The vertical cross-sections correspond to a family of observers at relative rest as they pass through the equator, the horizontal cross-sections to the family of Cauchy surfaces orthogonal to these observers' worldlines. By their lights, space collapses down to a minimum size, then expands.

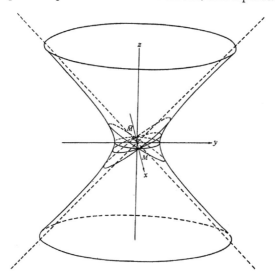

Figure 1.6 Boosting the equator in the y direction results in a family of surfaces that intersect at the points M and $\overline{M}$ (these will be spheres in higher dimensions). (This diagram is reproduced, by permission of Cambridge University Press, from Schrödinger, *Expanding Universes*, 15.)

In stationary spacetimes, time translation is a global symmetry. The existence of such a symmetry has momentous consequences. In generic curved spacetimes, there is no sensible notion of conserved total energy for a matter field—but this notion makes sense in stationary spacetimes.[17] In stationary spacetimes, the time translation symmetry plays a crucial role in singling out a natural notion of particle for quantum field theories—but there is no such distinguished notion in generic curved spacetimes.[18] These good things are unavailable in de Sitter spacetime.

> Although de Sitter space has the same number of Killing vector fields as Minkowski space, it has no analog of time translations, i.e. does not possess a Killing vector that is timelike everywhere. While this does not indicate any kind of causal pathology, it means that one cannot form a positive definite conserved Hamiltonian (energy) from the conserved stress tensor $T_{\mu\nu}$ of the Klein-Gordon field, or similar other fields.... In quantum field theory on de Sitter spacetime, it also makes it impossible to define a reasonable global notion of particle with similar properties as in Minkowski spacetime.[19]

4. The Conformal Completion

The Poincaré disk is a conformal completion of the hyperbolic plane: there is an angle-preserving map from the latter to the former that sends hyperbolic geodesics to finite curves whose endpoints correspond to "points at infinity" in hyperbolic geometry.[20] Start with the infinite hyperbolic plane: the non-Euclidean geometry of Bolyai and Lobachevsky. Choose an arbitrary point. Impose a rule according to which metre sticks grow as one moves outwards from this point in a cleverly chosen way. The result is that the entire infinite hyperbolic plane can be thought of as having finite extent as measured

---

with the same asymptotics as de Sitter spacetime is stationary, see Kesavan, *Asymptotic Structure of Space-Time with a Positive Cosmological Constant*, §4.2.1.

[17] On this point, see, e.g., Carroll, *Spacetime and Geometry*, §§3.5 and 3.8.

[18] See Wald, *Quantum Field Theory in Curved Spacetimes and Black Hole Thermodynamics*, chapter 4. For further discussion, see Ruetsche, *Interpreting Quantum Theories*, chapters 9 and 10.

[19] Hollands, "Correlators, Feynman Diagrams, and Quantum No-Hair in de Sitter Space-time," 6.

[20] For details concerning this construction, see, e.g., Ratcliffe, *Foundations of Hyperbolic Manifolds*, §4.5.

Figure 1.7 The Poincaré Disk.

by these stretchy metre sticks. Indeed there is a distance-preserving map from the hyperbolic plane equipped with its new funny metric structure to a finite Euclidean disk. This mapping distorts distances but preserves angles. The boundary of the disk can be thought of as consisting of ideal points added to the hyperbolic plane to represent points at infinity (i.e., directions of hyper-parallel lines). Figure 1.7 represents the result of first tessellating the hyperbolic plane by congruent triangles, then compactifying to the Poincaré disk—the angles of the triangles are preserved while distances are massively distorted. In effect, we are exploiting the fact that we can think of the disk as equipped with two different geometric structures—as a region of the Euclidean plane of finite extent or as the entire infinitely extended hyperbolic plane—with everything set up in such a way that these two geometric structures agree about facts about angles.

We would like to do the same sort of thing with de Sitter spacetime: choose a rule that distorts space and time measurements in a way that allows us to map the entire temporally infinite dS_d into a compact region of some other spacetime (whose boundaries can be thought of as corresponding to past and future infinity for de Sitter spacetime), even while preserving light cone structure.

We begin by recalling some facts about the d-dimensional Einstein static universe, E_d. This is a stationary solution with topology $\mathbb{R} \times S^{d-1}$ with metric of the form $ds^2 = -dt^2 + d\Omega_{d-1}^2$, where $d\Omega_{d-1}^2$ is the line element of S_{d-1}.[21] This geometry in effect has both absolute space and absolute time: the (connected component of the identity of the) isometry group is a product of a time translation subgroup and a six-dimensional subgroup consisting of spatial rotations; the orbits of the former subgroup can be thought of as the points of absolute space; the orbits of the latter subgroup are orthogonal to these privileged worldlines and can be thought of as the instants of absolute time.[22] E_d can be pictured as an infinite d-dimensional cylinder: the time translation symmetries move points along the direction parallel to the cylinder's axis; the orthogonal spatial sections are round spheres S_{d-1}.

Let us choose two instants of absolute time, $\mathcal{I}^-$ and $\mathcal{I}^+$ in E_d, with the temporal separation Δ between them being chosen so that if two photons are fired in opposite directions from any point on $\mathcal{I}^-$ they will meet again for the first time at a point on $\mathcal{I}^+$. Let us use $E_d(\Delta)$ to denote the region of E_d lying between $\mathcal{I}^-$ and $\mathcal{I}^+$ (and including its boundaries). Figures 1.8 and 1.9 show $E_d(\Delta)$, first as a cylinder and then unrolled as a rectangle (if $d > 2$, then some spatial dimensions are being suppressed, of course).[23]

It is possible to map all of dS_d onto $E_d(\Delta)$ in a way that distorts measurements of space and time but which preserves lightcone structure.[24] So we can use the lightcone structure of $E_d(\Delta)$ to study dS_d. For this purpose, it is convenient to think of dS_d as the subset of $E_d(\Delta)$ lying (strictly) between $\mathcal{I}^-$ and $\mathcal{I}^+$ (so we think of this region as having both a de Sitter metric and an Einstein static metric—with everything set up so that these two metrics agree about which curves represent paths of photons). From the de

---

[21] More generally, we could consider $E_d(r)$, the Einstein static universe in dimension d whose natural spatial sections are spheres of radius r.

[22] Einstein on his static universe, in a postcard to Ehrenfest on Valentine's day 1917 (8.298): "The odd thing is that now a quasi-absolute time and a preferred coordinate system do reappear in the end, while fully complying with all the requirements of relativity."

[23] There are varying conventions for such diagrams—but it is more common to depict the result of unrolling $E_d(\Delta)$ as a square. Here we want to emphasize that photons sent in opposite directions from $\mathcal{I}^-$ meet at a point on $\mathcal{I}^+$, so we proceed as in Griffiths and Podolský, *Exact Space-Times*, §4.4.1.

[24] That is: there is a conformal isometry between dS_d and $E_d(\Delta)$ (recall that a *conformal isometry* between (M, g) into (N, h) is a diffeomorphism $d : M \to N$ such that $d^*g = \omega^2 h$ for some $\omega : N \to \mathbb{R}^+$). See fn. 8 of Chapter 2 below for an explicit example of such a conformal isometry between dS_d and $E_d(\Delta)$.

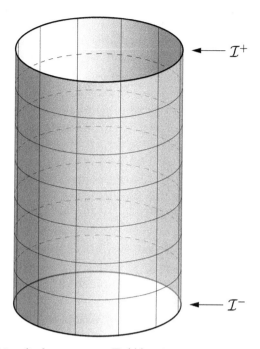

Figure 1.8 This cylinder represents $E_d(\Delta)$, a finite portion of the Einstein static universe with the feature that a photon can travel halfway around the universe between the beginning of time at $\mathcal{I}^-$ and the end of time at $\mathcal{I}^+$. This picture is accurate if $d = 2$ (so that spatial sections are circles); for $d = 4$, each point represents a two-sphere (and spatial sections are three-spheres).

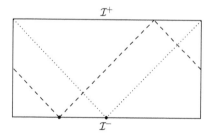

Figure 1.9 $E_d(\Delta)$ unrolled as a rectangle. The dotted lines represent two photons sent in opposite directions from a point on $\mathcal{I}^-$, which meet again for the first time at $\mathcal{I}^+$. Likewise for the dashed lines. In both cases, it is important to remember the rectangle represents a cylinder—opposite points on the vertical edges are identical.

Sitter perspective, points on the surfaces $\mathcal{I}^-$ and $\mathcal{I}^+$ are ideal points—the "endpoints" of infinitely extended timelike or null geodesics.

Let (M, g) be any time-oriented Lorentz manifold and let ℓ be any timelike curve in (M, g). Recall that the *causal future*, $J^+(\ell)$, of ℓ is the set of spacetime points y such that there exists a point x on ℓ such that there is a future-directed causal curve connecting x to y, so that a massive particle or a photon can be sent from x to y. Dually, the *causal past*, $J^-(\ell)$, of ℓ is the set of spacetime points x such that there exists a point y on ℓ such that there is a future-directed causal curve connecting x to y. The *causal diamond* of ℓ is the intersection of the causal past and causal future of ℓ—the set of spacetime points that can both signal and be signalled by ℓ.

Figures 1.10 and 1.11 are *Penrose diagrams* of de Sitter spacetime: spacetime diagrams that exploit a conformal completion. They illustrate the concepts just defined. Roughly and heuristically speaking, for any eternal worldline ℓ in dS, $J^-(\ell)$ and $J^+(\ell)$ each cover one-half of spacetime—which half depending only on the past/future endpoint of ℓ on $\mathcal{I}^\pm$. The causal diamond of ℓ covers a smaller region that depends on both the future and past endpoints of ℓ. So de Sitter spacetimes feature eternal observer horizons: for any observer, there are regions of spacetime that that observer cannot signal and regions of spacetime from which that observer cannot be signalled. The geometries of the causal futures and causal diamonds of de Sitter observers will play a prominent role in our investigation of the nature of time and simultaneity of de Sitter spacetime in the next chapter.

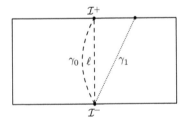 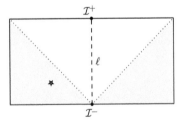

Figure 1.10 On the left: three worldlines asymptotic to each other in the past (and hence sharing the same endpoint on $\mathcal{I}^-$); two of these are also asymptotic to each other in the future. On the right: the light portion represents the causal future $J^+(\ell)$ of the worldline ℓ—the region to which signals can be sent from points on ℓ. The star represents an event that can signal but not be signalled by points on ℓ.

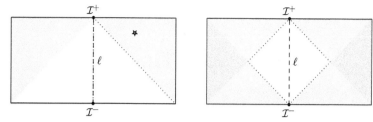

Figure 1.11 On the left: the causal past $J^-(\ell)$—the region from which signals can be sent to points on ℓ. The star represents an event that can be signalled by but which cannot signal points on ℓ. On the right: the causal diamond of ℓ. The white region consists of points that can signal and be signalled by ℓ. The darker shaded region consists of points that can neither signal nor be signalled by ℓ. The lighter shaded region consists of points that can either signal ℓ or be signalled by ℓ, but not both.

EXERCISE 1.7 (Medium). Consider the Einstein static universe E_d ($d \geq 2$). Let γ_0 be one of the geometrically privileged timelike geodesics in E_d. Let p and q be points on γ_0 separated by $\tau > 0$ units of proper time.

a) Let $\kappa(d,\tau)$ be the cardinality of the set of timelike geodesics segments that have p and q as endpoints. Determine how κ depends on d and on τ.

b) Let γ_1 be a timelike curve that passes through both p and q and which does not coincide with γ_0 between p and q. Show that the amount of proper time that elapses between p and q along γ_1 is strictly less than τ.[25]

[25] HINT: it may be helpful to have in mind the following alternative characterization of the Einstein static universe. Let t, x_1, x_2, ..., x_d be inertial coordinates for $(d+1)$-dimensional Minkowski spacetime. Then E_d is the result of restricting the Minkowski metric to the locus $x_1^2 + x_2^2 + \ldots + x_d^2 = 1$.

2
Let the Good Times Roll

1. Introduction

In Minkowski spacetime, there is of course no absolute notion of simultaneity. But specifying a timelike geodesic ℓ_0 in Minkowski spacetime (the worldline of a freely falling observer) determines a geometrically unique foliation of $\mathbb{M}$ by flat hypersurfaces, the surfaces of simultaneity relative to ℓ. Each of the following three constructions leads to the same result.[1]

- i) OPERATIONAL: let ℓ_0 employ Einstein's simultaneity convention.
- ii) SYMMETRY: find the flat spacelike hypersurfaces orthogonal to ℓ_0.
- iii) REST FRAME: find the spacelike hypersurfaces orthogonal to the family of observers comoving with ℓ_0.

Things work quite differently in de Sitter spacetime. The upshot is summarized by Einstein in a postcard to Klein. Misled by a coordinate singularity, Einstein had initially denied that de Sitter's solution was a homogeneous and non-singular $\Lambda > 0$ vacuum solution. He eventually conceded the point, but immediately raised another objection, this time based on a sound geometrical understanding.

Esteemed Colleague,
You are entirely right. De Sitter's world is, in and of itself, free of singularities and its space-time points are all equivalent. . . . However, under no condition could this world come into consideration as a physical possibility. For in this world, time t cannot be defined in such a way that the three-dimensional slices $t =$const. do not intersect one another and so that these slices are equal to one another (metrically).[2]

---

[1] For a helpful overview, see Giulini, "Uniqueness of Simultaneity."
[2] Postcard to Klein of 20 June 1918 (8.567).

Accelerating Expansion: Philosophy and Physics with a Positive Cosmological Constant. Gordon Belot, Oxford University Press. © Gordon Belot 2023. DOI: 10.1093/oso/9780192866462.003.0003

The next four sections below will unpack Einstein's claims here, exploring the peculiarities of several different notions of de Sitter simultaneity. We will see that some interesting slicings of de Sitter spacetime into instants of time cover only part of the de Sitter hyperboloid—these slicings and the coordinates adapted to them will play an important role in subsequent chapters. In the final section of the chapter, we consider the sense in which de Sitter geometry requires a revision of the pre-relativistic notion of time yet deeper than that required by Minkowski geometry.

2. Einstein Simultaneity and the Static Patch

Recall the Einstein simultaneity convention for a freely falling observer with worldline ℓ in a relativistic spacetime. Each point on ℓ is simultaneous with itself and with no other point on ℓ. And if a light signal emitted at event p_1 on ℓ would be received and reflected by event r and return to ℓ at event p_2, then ℓ considers r to be simultaneous with the event p on ℓ that occurs halfway in proper time between p_1 and p_2. In this way, freely falling observers are able to consider distant events to be simultaneous with events on their worldlines. But since only events that can both signal and be signalled by events on ℓ can be considered to be Einstein-simultaneous with points on ℓ, only events in the causal diamond $J^+(\ell) \cap J^-(\ell)$ of ℓ will be eligible for this honour.

In Minkowski spacetime, all events are in the causal diamond of the worldline of an eternal freely falling observer and Einstein-simultaneity is an equivalence relation. The picture is different in de Sitter spacetime. The causal diamond of a freely falling observer covers only a fraction of the spacetime (see Figure 2.1)—and since only points in the interior of the causal diamond of an observer can be Einstein-simultaneous with points on the observer's worldline, the surfaces of Einstein simultaneity of a de Sitter observer do not partition de Sitter spacetime.

Within the causal diamond of a freely falling de Sitter observer ℓ_0, what do the equivalence classes of Einstein simultaneity look like? Consider again the de Sitter hyperboloid dS_2 as a subset of $\mathbb{M}_3$.

Figure 2.1 shows the causal diamond of the timelike geodesic γ_0 that is the $y > 0$ branch of the intersection of dS_2 with the yz-plane. Note that only the right half of the equator as pictured in this diagram lies in the interior of the causal diamond of γ_0. Note also that the right half of the equator includes a single point of γ_0, which we will call y^+. As one would expect, every point in the right half of the equator is Einstein-simultaneous with y^+ (relative

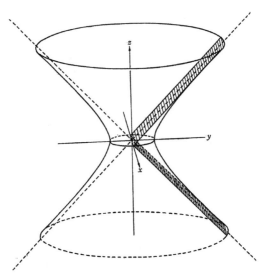

Figure 2.1 The causal diamond of an observer moving along the $+y$ branch of the intersection of the plane $x = 0$ with the de Sitter hyperboloid. The causal diamond includes half of the equator and is bounded by null geodesic segments departing from the endpoints of this half-equator. (This diagram is reproduced, by permission of Cambridge University Press, from Schrödinger, *Expanding Universes*, 19.)

to γ_0)—and no other point in dS_2 is (see the exercises below). Any point y^* on γ_0 can be reached by acting on y^+ by a de Sitter symmetry corresponding to a suitable boost in the y direction in the ambient $\mathbb{M}_3$. And of course the set of points Einstein-simultaneous (relative to γ_0) with such a y^* will just be the image under the relevant boost of the right half of the equator (Figure 1.6 depicts the result of acting on the equator of dS_2 by boosts in the y direction). Note that the xy-plane and its various boosts in the y direction all intersect along the x-axis—the equator and its boosts all intersect at the points labelled M and $\overline{M}$ in Figure 1.6—but these points lie on the boundary of the causal diamond of γ_0 and so are not Einstein-simultaneous with points on γ_0. All of this carries over nicely to higher-dimensional de Sitter spacetimes (although in higher dimensions, M and $\overline{M}$ will be spheres of appropriate dimension).

The interior of the causal diamond of a timelike geodesic γ of dS_d, depicted in Figure 2.2, is often called the *static patch* of γ because it can

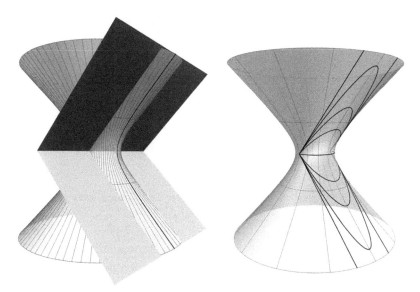

Figure 2.2 Two renderings of the static patch of a timelike geodesic in dS_2. On the left: the behaviour of geodesics comoving through the equator with the geodesic the patch is centred on. On the right: some surfaces of constant t in the static patch coordinates (the null segments the delineate the boundary of the patch can be thought of as corresponding to $t = \pm\infty$).

be given coordinates in which the components of the metric are time-independent:

$$ds^2 = -(1-\rho^2)dt^2 + (1-\rho^2)^{-1}d\rho^2 + \rho^2 d\Omega_{d-2}^2 \quad \text{Static Patch Metric}$$

with $\rho \in [0,1)$, $t \in \mathbb{R}$, $d\Omega_{d-2}^2$ being the line element for $(d-2)$-dimensional round sphere, and with the level surfaces of t being the hypersurfaces of Einstein simultaneity of the observer γ at rest at $\rho = 0$. This metric is *static* (= stationary and time-reversal symmetric) with ∂_t being the generator of time translation. Note that on the boundary of the static patch (given by $\rho = 1$), this expression for the de Sitter metric becomes ill-defined and ∂_t ceases to be timelike.[3] Away from the equator, points near the boundary

---

[3] This is the coordinate singularity that initially led Einstein astray in his interpretation of de Sitter's solution.

of the static patch are assigned arbitrarily late times by γ_0—even though it would be easy for an observer whose worldline initially coincided with γ_0 to accelerate slightly and then to escape from the static patch in a finite amount of proper time (this may remind you of the behaviour of the event horizon in Schwarzschild spacetime).

The static patch is, typically, eventually a very lonely place: the only way for another timelike geodesic to remain in the static patch of a given timelike geodesic towards the future is for the two to be asymptotic towards the future (and likewise for the past).

EXERCISE 2.1 (Easier). Why do we restrict the Einstein simultaneity convention in Minkowski spacetime to inertial observers?

a) Show that if a uniformly accelerating observer in Minkowski spacetime applies the Einstein simultaneity convention, the resulting hyperplanes of simultaneity do not form a partition of Minkowski spacetime.

b) Argue that if any accelerating observer in Minkowski spacetime applies the Einstein simultaneity convention, the resulting hyperplanes of simultaneity do not form a partition of Minkowski spacetime.

EXERCISE 2.2 (Medium). Fix inertial coordinates (t, x, y) in $\mathbb{M}_3$ and consider dS_2 as a subspace of $\mathbb{M}_3$ in the usual way. Let γ be the $y > 0$ half of the intersection of dS_2 with the $x = 0$ plane in $\mathbb{M}_3$. Let R be the subset of the $t = 0$ equator of dS_2 that satisfy $y > 0$. Let p be the point of intersection between γ and R.

a) Show that γ considers every point in R to be simultaneous with p.

b) Show that γ does not consider any point in R to be simultaneous with any point on γ other than p.

c) Show that γ does not consider any point not in R to be simultaneous with p.

EXERCISE 2.3 (Medium). Consider the static patch of some inextendible timelike geodesic γ_0 in the coordinates given above. Using the geodesic equation, show that aside from γ_0 itself, integral curves of the vector field ∂_t are not geodesics. Show that among geodesics comoving with γ_0 through the equator, only γ_0 itself remains permanently in the static patch towards the future (or the past). Show that the only way that a timelike geodesic can remain permanently in the static patch of γ_0 towards the future is to be asymptotic towards the future to γ_0.

EXERCISE 2.4 (Medium). Let γ_0 be an inextendible timelike geodesic in de Sitter spacetime. Construct a second such geodesic, γ, asymptotic to γ_0 towards the past. Show that γ is not asymptotic to γ_0 towards the future.[4]

3. Simultaneity via Symmetry

Let ℓ be an inertial worldline in Minkowski spacetime and let p be a point on ℓ. The flat hypersurface Σ_p orthogonal to ℓ at p is invariant (as a set) under all of the symmetries of Minkowski spacetime that fix p and leave ℓ invariant as a set. Further, it is the only Cauchy surface with this invariance property (since time-reflection is among the symmetries in question).

We can do something similar in the de Sitter setting (see Figure 2.3). Let γ be an inextendible timelike geodesic in de Sitter spacetime and let q be a point on γ. There is a unique flat hyperplane Π_0 of the ambient Minkowski spacetime that passes through the origin and which is orthogonal to γ at q. So the de Sitter Cauchy surface Σ_0 that arises as the intersection of the de Sitter hyperboloid with Π_0 is definable in terms of q and γ and hence is invariant under every de Sitter symmetry that fixes q and leaves γ invariant as a set. Σ_0 is of course just the equator of the hyperboloid relative to inertial coordinates according to which Π_0 is given by the condition $t = 0$. Further: relative to such coordinates, the transformation $t \mapsto -t$ is a Lorentz symmetry that fixes q and leaves γ invariant as a set—so any Cauchy surface invariant under all such symmetries must be a subset of (and hence coincide with) Σ_0. And, of course, if p is some other point on γ, then p will be related to q by a boost in a certain direction (determined by γ) and the unique de Sitter Cauchy surface through p invariant under all of the relevant symmetries will be the image of Σ_0 under this boost. The family of all of the Cauchy surfaces that determined in this way by points on γ will not partition de Sitter spacetime: they will not be disjoint (they will share certain points—see Figure 2.3); and they will not include any points that are not in the static patch of γ or the static patch of $-\gamma$ (the family of points antipodal to γ).

---

[4] HINT: note that γ_0 is asymptotic towards the past to some null geodesic ℓ of the ambient Minkowski spacetime.

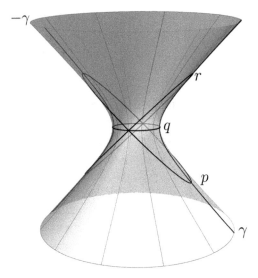

Figure 2.3 Points p, q, and r on timelike geodesic γ. The hypersurface through q depicted here is the unique hypersurface invariant under symmetries of the hyperboloid that fix q and leave γ invariant as a set. Likewise, *mutatis mutandis*, for the hypersurfaces through p and r depicted here.

4. Simultaneity via Comoving Observers—Flat Slices

Here is yet another procedure that leads from the specification of a freely falling observer in Minkowski spacetime to the hypersurfaces of Einstein simultaneity for that observer: given a timelike geodesic ℓ in Minkowski spacetime, construct the family of observers that remain at constant distance from ℓ, then construct the family of spacelike hypersurfaces everywhere orthogonal to this family of observers. This construction cannot be carried over to the de Sitter setting: Λ tends to push initially nearby observers further and further apart from one another, so it is not possible to construct a family of observers that maintain their distances from a given freely falling observer.[5]

But here is another, more elaborate, version of the construction for Minkowski spacetime: (i) given a timelike geodesic ℓ and a point p on ℓ, construct the flat spacelike hypersurface Σ_p that is orthogonal to ℓ at p; (ii) at each point q on Σ_p, construct the timelike geodesic ℓ_q orthogonal to Σ_p at q; (iii) construct the family of spacelike hypersurfaces everywhere orthogonal

---

[5] Compare: in Euclidean space, it makes sense to ask for the family of lines parallel to a given line—but on the surface of a sphere, it is not possible to find even one great circle that remains parallel to a given great circle.

to the ℓ_q. The resulting hypersurfaces are flat and partition Minkowski spacetime—they are, of course, surfaces of Einstein simultaneity of ℓ.

Remarkably, although dS has the topology $\mathbb{R} \times S^3$, a version of this recipe can be executed in the de Sitter setting. Let γ be the worldline of an eternal freely falling observer in de Sitter spacetime. At any point p on γ, it is possible to find a flat spacelike de Sitter hypersurface Σ orthogonal to γ at p and that is contained in $J^+(\gamma)$.[6] One can then find the family of timelike geodesics orthogonal to this surface and the family of hypersurfaces everywhere orthogonal to these geodesics. These hypersurfaces cannot be Cauchy surfaces (in dS each Cauchy surface has the topology of a three-sphere). And they partition only the causal future of γ—see Figure 2.4. The family of slices constructed depends on γ but not on the choice of p.

The de Sitter metric on $J^+(\gamma)$ can be written in the Robertson-Walker form

$$ds^2 = -dt^2 + e^{2t}(dx_1^2 + dx_2^2 + \ldots + dx_{d-1}^2) \qquad \text{The Cosmological Patch}$$

with, t and the x_i ranging over all of $\mathbb{R}$, the level surfaces of t being our flat hypersurfaces, and t again measuring proper time along our special timelike geodesics (see Figures 2.5 and 2.6).[7]

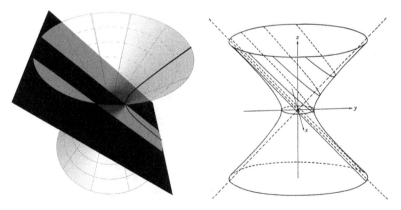

Figure 2.4 Two renderings of the cosmological patch of dS. (The diagram on the left is reproduced, by permission of Cambridge University Press, from Schrödinger, *Expanding Universes*, 29.)

---

[6] Remarkably, such surfaces arise as the intersection of the de Sitter hyperboloid with certain null hypersurfaces of the ambient Minkowski spacetime—see Schrödinger, *Expanding Universes*, 29 f.
[7] Recall that a Robertson-Walker metric is of the form $ds^2 = dt^2 + a(t)d\Sigma^2$ where $d\Sigma^2$ is a Riemannian metric of constant sectional curvature.

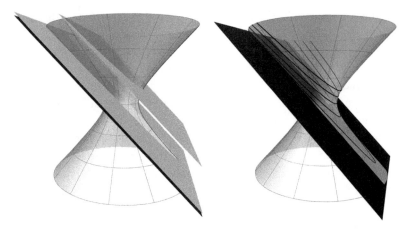

Figure 2.5 Each level surface of t in the cosmological patch arises as the intersection of the hyperboloid with a null hypersurface parallel to the null hypersurface that defines the patch. On the left, the defining null hypersurface (dark) and two parallel to it (light). On the right, the corresponding level surfaces of t (along with three more intermediate between them).

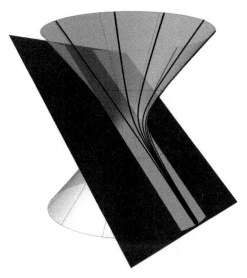

Figure 2.6 Five timelike geodesics in the cosmological patch that arise by fixing values of the spatial coordinates and allowing t to vary.

5. Simultaneity via Comoving Observers—Spherical Slices

Let us reconceive yet again the recipe that takes us from a freely falling observer ℓ in Minkowski spacetime to the surfaces of Einstein simultaneity of ℓ via the family of observers comoving with ℓ. We can describe the recipe as follows: (i) given ℓ and a point p on ℓ, find a geometrically distinguished spacelike slice Σ_p orthogonal to ℓ at p; (ii) construct the family of timelike geodesics orthogonal to Σ_p; (iii) construct the family of spacelike hypersurfaces orthogonal to this family of geodesics.

This version of the construction differs from the previous one only by allowing us to use any type of geometrically privileged hypersurface at the first stage, rather than requiring us to employ flat hyperplanes. Here is a natural implementation of this generalized strategy in de Sitter spacetime: choose a timelike geodesic γ and a point p on γ; let Σ_p be the unique Cauchy surface of minimum volume orthogonal to γ at p; construct the family of timelike geodesics orthogonal to Σ_p (the family of observers whose worldlines are orthogonal to the equator, relative to the coordinates just mentioned); and construct the family of spacelike hypersurfaces everywhere orthogonal to this family. The result will be a family of hypersurfaces that can be thought of as the level surfaces of t relative to a system of inertial coordinates on the ambient Minkowski spacetime (see Figure 2.7).

Figure 2.7 The level surfaces of t in the global patch coordinates.

In any dimension, no matter what timelike geodesic and what point on it we take as input, the first Cauchy surface we construct via this procedure will be the only one of minimum volume in the resulting family. The family of timelike geodesics we construct can be thought of as comoving observers in a Robertson-Walker spacetime who see space as having arbitrarily large volume in the past, shrinking exponentially fast down to minimum volume, then expanding exponentially. The surfaces of simultaneity relative to these observers are surfaces of constant t when we write the de Sitter metric in the form:

$$ds^2 = -dt^2 + \cosh^2 t \, d\Omega_{d-1}^2 \quad \text{The Global Patch}$$

with $t \in \mathbb{R}$ and $d\Omega_{d-1}^2$ the line element of the round $(d-1)$-sphere.[8]

6. Flow and de Sitter

6.1 Newtonian Physics

Newton, of course believed that simultaneity was absolute. What is surprising is that he explicitly tells us so:

> we do not ascribe various durations to the different parts of space, but say that all endure simultaneously. The moment of duration is the same at Rome and at London, on earth and on the stars, and throughout all the heavens.... [W]e understand any moment of duration to be diffused throughout all spaces, according to its kind.[9]

For Newton, time consists of a continuous family of moments of duration, occurring successively. And because time has this structure it makes sense fall back upon the ancient metaphor of flow:

---

[8] We can now exhibit a conformal isometry between dS_d and our temporally finite portion $E_d(\Delta)$ of the Einstein static universe. Making the coordinate transformation $t \mapsto \tau = 2\arctan e^t - \frac{\pi}{2}$, the de Sitter metric becomes $\omega^{-2}(-d\tau^2 + +d\Omega_{d-1}^2)$ where $\omega = \cosh^{-1}\tau$ and $-\frac{\pi}{2} < \tau < \frac{\pi}{2}$. In other words: we have exhibited the de Sitter metric as a positive multiple of the metric of Einstein's static universe.

[9] From Newton's manuscript *De Gravitatione*. Translation of Janiak (ed.), *Newton: Philosophical Writings*, 26.

Absolute, true, and mathematical time, in and of itself and of its own nature, without reference to anything external, flows uniformly and by another name is called duration.[10]

In unpublished notes, Einstein sums up the Newtonian account as follows:

Ask an intelligent man who is not a scholar what space and time are, and he will perhaps answer as follows. If we imagine all physical things, all stars, all light taken out of the universe, what then remains is something like a giant vessel without walls called "space." With respect to what is happening in the world, it plays the same role as the stage in a theater performance. In this space, in this vessel without walls, there is an eternally uniformly occurring tick-tock that, however, only ghosts can hear, but those everywhere; that is "time." Most natural scientists, up to the present, had this conception about the essence of space and time, even though they did not phrase it in such naive terms as we just did for the sake of simplicity.[11]

6.2 Special Relativity

Following the passage just quoted, Einstein of course argues that absolute simultaneity cannot be consistently combined with the Galilean principle of relativity and the light postulate (the thesis that the speed of light, unlike the speed of bullets or of water waves, is independent of the frame of emission or of measurement).

There is no audible tick-tock everywhere in the world that could be considered as time. If physics wants to use time, it first has to define it. In this endeavor it is apparent that this definition necessarily requires a body of reference. ... It turns out that one can define time relative to this body of reference such that the law of the propagation of light is obeyed relative to it. ... But it turns out that the times of differently moving bodies of reference do not coincide. ... If two events occurring at different locations are judged simultaneous from a body of reference, then they are not judged so from a body of reference that is moving relative to it.

---

[10] From the Scholium to the Definitions in book one of Newton's *Principia Mathematica*. Translation of Cohen and Whitman (eds.), *Isaac Newton: The Principia*, 408.

[11] Einstein, "The Principle Ideas of the Theory of Relativity" (6.44a). Translation emended.

For more than a century, Einstein's abolition of absolute simultaneity has bedevilled philosophers interested in making sense of the passage or flow of time. If time flows, then it would appear that there must be an absolute fact about whether A and B occur at the same time, or A precedes B, or vice versa—but Einstein says that when A and B are spacelike separated, there can be no such fact.

Of course, this is by no means the end of the story. Although the geometry of Minkowski spacetime will not allow one to define an invariant standard of simultaneity, if this geometry is supplemented by a choice of an inertial frame, the resulting structure *does* allow one to define an absolute notion of simultaneity. Indeed, the structure that results from adding a distinguished inertial frame to the structure of Minkowski spacetime is essentially just Newtonian spacetime (plus the choice of a special absolute velocity, the speed of light).

The most widely pursued route to constructing a post-Einsteinian account of the flow or passage of time is to deny that Minkowksi geometry is the whole story about spatiotemporal geometry, and to maintain that the whole story is something more like Minkowski geometry supplemented by a choice of a preferred frame.[12] And now it is no harder (nor easier) to make sense of the passage of time than it was in the Newtonian regime.

Such approaches are unapologetically revisionary: they require us to think that Einstein misled us by convincing us to take the Galilean principle of relativity too seriously—it is a mistake to take all inertial frames to be equivalent. There is a subtler way to go: one can maintain that there are many flowing times, one for each of the inertial frames of Minkowski spacetime.[13] Since all frames are treated as on a par, there is no need to call for a counter-revolution to overthrow special relativity. But compared with the preceding type of approach, this requires taking on an extra burden: in addition to making sense of the flow of time within Newtonian spacetime, one also has to make sense of the idea that in a special relativistic world there are many Newtonian realities, each with its own flowing time.[14]

---

[12] See, e.g., Tooley, "spatiotemporal Defense of Absolute Simultaneity."
[13] For an approach of this kind, see Fine, "Tense and Reality," §10.
[14] There are of course yet other ways to attempt to combine passage/flow with special relativity—see, e.g., Skow, *Objective Becoming*, chapters 8 and 9.

6.3 General Relativity

The move to general relativity brings with it strange new temporal be-haviours. One the one hand, some relativistic cosmologies support geo-metrically privileged candidates for a relation of absolute simultaneity—suggesting that the special relativistic revolution should be rolled back.[15] On the other hand, time is cyclic for all observers in some general relativistic worlds (such as the anti-de Sitter hyperboloid) and in others, time is cyclic for some observers and linear for others (as in Gödel spacetime)—suggesting that there is no going back.[16]

The lesson of our exploration of simultaneity in de Sitter spacetime is that the transition from Minkowski spacetime to de Sitter spacetime opens a new front. This transition is in many ways just as dramatic as the transition from Newtonian spacetime to Minkowski spacetime. In Newtonian physics, the ancient metaphors of passage and flow find a comfortable home in Newtonian spacetime. In Minkowski spacetime, there is no absolute time: simultaneity itself is relative to a choice of observer or frame. But if one supplements the geometry of Minkowski spacetime by a choice of a preferred observer or frame, one essentially reinstates the Newtonian picture. So we can think of Minkowski spacetime as in effect containing many copies of Newtonian spacetime laid askew to one another (one for each inertial frame). Not so de Sitter spacetime: relative to a choice of observer or frame, one can construct notions of simultaneity—but either they do not cover all of spacetime or they do not respect its homogeneity. The fact that de Sitter spacetime is homogeneous but not stationary is a sign that we are encoun-tering something new again, something to which our inherited temporal concepts are not well-adapted. In particular, our old metaphors no longer fit comfortably. In de Sitter spacetime, time rolls along rather than flowing: the closest thing there is to a notion of time translation (as in Newtonian spacetime) or a family of such notions (as in Minkowski spacetime) are the families of hyperbolic rotations induced by Lorentz boosts of the ambient Minkowski spacetime.

---

[15] For early advocacy of this stance, see Eddington, *Space, Time, and Gravitation*, 163, and Jeans, "Man and the Universe," 21 f. For a recent overview of the dialectic, see Zimmerman, "Presentism and the Space-Time Manifold," §4.6.

[16] See Gödel, "A Remark about the Relationship between Relativity Theory and Idealistic Philosophy"—this paper was in part a reaction to the works of Eddington and Jeans cited in the preceding note. In parallel to the argument from time travel, this paper features an intriguing symmetry argument—for discussion, see Belot, "Dust, Time, and Symmetry."

QUESTION 2.1 (Smaller). General relativity predicts (and experiment confirms) gravitational time dilation.

a) Read Einstein's 1911 paper "On the Influence of Gravitation on the Propagation of Light" (3.23) and extract from it a simple, convincing thought experiment showing that clocks should run more slowly the deeper they are in a gravitational potential well. What physical assumptions are required?

b) Rovelli makes the following remarks regarding flow and gravitational time dilation:

> Clocks may well run at different speeds in the mountains and in the plains, but is this really what concerns us, ultimately, about time? In a river, the water flows more slowly near its banks, faster in the middle—but it is still flowing.... Is time not also something that always flows—from the past to the future?[17]

How deep a revision of our ordinary notions of flow and passage are required in order to accommodate gravitational time dilation?

QUESTION 2.2 (Larger). When Newton said that time flowed equably, there are two things that he probably had in mind: something about the chronogeometric structure of our world (it is invariant under time translation) and something about the laws of nature (they are invariant under time translation). Gödel thinks that facts about the chronogeometric structure of our world, taken on their own, would be insufficient to validate the claim that time flows or that change is objective—he thinks such a claim is in some sense a claim about the full range of relativistic worlds. Is he right?[18]

EXERCISE 2.5 (Easier). Another crucial general relativistic novelty: the role of global spacetime structure. Here is a classic way to illustrate this aspect of the theory. Consider two spacetimes that have the same local geometry as Minkowski spacetime but which have different global structures.

i) Begin with $\mathbb{R}^2 = \{(t, x) \mid t, x \in \mathbb{R}\}$, equipped with the standard Minkowski metric. Choose real numbers $a < b$ and delete all points with x coordinates less than a or greater than b, then identify points of

---

[17] Rovelli, *The Order of Time*, 19 f.
[18] For inconclusive discussions of this issue, see Earman, *Bangs, Crunches, Whimpers, and Shrieks*, appendix to chapter 6; and Belot, "Dust, Time, and Symmetry," §3.

the form (t, a) and (t, b) to yield a cylinder with a flat Lorentz metric, in which surfaces of constant t are spacelike and have topology S^1.

ii) Again begin with $\mathbb{R}^2 = \{(t, x) \mid t, x \in \mathbb{R}\}$ with the Minkowski metric and choose $a < b$. This time delete all points whose t coordinate is less then a or greater than b and identify points of the form (a, x) with points of the form (b, x) to yield a flat cylindrical spacetime in which surfaces of constant t are spacelike and have the topology of $\mathbb{R}$.

Show that both of these cylindrical spacetimes admit geometrically privileged notions of simultaneity. What sorts of experiments could inhabitants of these worlds perform to identify these absolute notions of simultaneity? How does the twin paradox look in these spacetimes?

QUESTION 2.3 (Larger). *Presentists* maintain that only present events exist—past and future events do not exist. They hasten to add that some logically contingent claims about the past and future are true and others false—and that it is facts about the present that determines which are which.[19] So presentists maintain that the present moment is related in a special way (via the laws of nature) to a one-parameter family of physically possible states.

a) Presentism requires symmetry breaking: in Newtonian physics, to select one instant from the one-dimensional continuum of possible instants; in special relativity, to select on instant from the four-dimensional continuum of flat spatial hyperplanes. Does the shift from Newtonian to special relativistic physics pose any special problem for presentists?

b) Let us say that *standard* presentists hold that while only the present moment is real *now*, other moments of time have been or will be real; while *magical moment* presentists hold that only one instant of time has occurred, is occurring, or will occur. Is there a real distinction here? If so, what are the relative advantages and disadvantages of these two views?

QUESTION 2.4 (Medium). In a passage inserted in a later German version of his paper on relativity and idealistic philosophy, Gödel remarks concerning his solutions with closed timelike curves:

---

[19] For a classic critical discussion of this claim about truthmakers, see Sider, *Four-Dimensionalism*, chapter 2.

The second law of thermodynamics would also seem to be compatible with the solutions above. For within them a positive direction can be defined for all time-like lines in a unique and continuous way. Furthermore, the probability of any material system returning *exactly* to a former state is vanishingly small; and if that happens only approximately, it merely means that somewhere two examples of the same system (in general having different entropies) exist simultaneously side by side.[20]

How do you understand Gödel's idea here? Is his interpretation of the second law the only plausible one?

---

[20] Gödel, "A Remark," 206.

3
Symmetry and Curvature

1. Introduction

The first two chapters above were concerned with de Sitter spacetime itself. Chapters 6–9 below focus on asymptotically de Sitter spacetimes, of central interest in cosmology, whose geometry may not anywhere look exactly like that of de Sitter spacetime. The ideal bridge between these two phases would be a thorough discussion of all Lorentz manifolds that have the local geometry of de Sitter spacetime—or, even better, a thorough discussion of all Lorentz manifolds of constant curvature. That is a large topic and one far, far beyond the reach of this work.[1] This chapter is modest in its ambitions: it will survey basic facts about (and examples of) Lorentz manifolds and Riemannian manifolds of constant curvature, focusing on those that exhibit a high degree of symmetry. The main goal is to introduce various notions and examples that will play a significant role in later chapters: various notions of symmetry, the highly symmetric elliptic variants of spherical and de Sitter spaces, and the hyperbolic and anti-de Sitter spaces that are the negative-curvature analogs of spherical and de Sitter spaces. Section 2 will treat the relevant notion of constant curvature (= constant sectional curvature). In Section 3, we will see a number precisifications of the informal notion of a highly symmetric space and delineate their relations to one another and to the condition of having constant curvature. Section 4 collects some facts about and examples of complete Riemannian manifolds of constant curvature. Section 5 does the same for Lorentz manifolds. (As this summary probably makes clear, this chapter is largely given over to technical exposition—some readers may prefer to skim parts of it, and to return later if and when they are so moved.)

---

[1] For the special case of complete Lorentz manifolds of constant curvature, see Calabi and Markus, "Relativistic Space Forms;" and Wolf, *Spaces of Constant Curvature*, §11.2. For the general case, see Mess, "Lorentz Spaces of Constant Curvature"; Andersson *et al.*, "Notes on: 'Lorentz Spaces of Constant Curvature'"; and Scannell, "Flat Conformal Structures and the Classification of de Sitter Manifolds."

Accelerating Expansion: Philosophy and Physics with a Positive Cosmological Constant. Gordon Belot, Oxford University Press. © Gordon Belot 2023. DOI: 10.1093/oso/9780192866462.003.0004

2. Spaces of Constant Curvature

Here we recall some basic facts about sectional curvature, including its relation to Riemannian curvature.[2]

Let (M, g) be an n-dimensional ($n \geq 2$) Riemannian manifold or Lorentz manifold.

Recall, first, the coordinate-based definition of the Riemann curvature tensor. Fix coordinate functions x^0, $x^1, \ldots x^{n-1}$ on (a neighbourhood of) M and use $[n]$ to denote the set $\{0, 1, 2, \ldots, n-1\}$. Relative to these coordinates, g is represented by an n-by-n matrix at each point x, the components of which we denote $g_{ab}(x)$ ($a, b \in [n]$). At each point x this matrix has an inverse, the components of which we denote $g^{ab}(x)$. For any $a, b, c \in [n]$, the *Christoffel symbol* of g is:

$$\Gamma^a{}_{bc}(x) := \frac{1}{2} \left[g^{ad}(x) \left(\partial_{x^c} g_{db}(x) + \partial_{x^b} g_{dc}(x) - \partial_{x^d} g_{bc}(x) \right) \right].$$

We build Riemann curvature tensor R of g out of the Christoffel symbols by specifying its components relative to our choice of coordinates. For $a, b, c, d \in [n]$:

$$\begin{aligned} R_{abc}{}^d(x) := {} & \partial_{x_b} \Gamma^d{}_{ac}(x) - \partial_{x_a} \Gamma^d{}_{bc}(x) \\ & + \sum_{e \in [n]} \left[\Gamma^e{}_{ac}(x) \Gamma^d{}_{eb}(x) - \Gamma^e{}_{bc}(x) \Gamma^d{}_{ea}(x) \right]. \end{aligned}$$

As usual, we use R_{abcd} to denote (the components of) the tensor that results from contracting the Riemann curvature tensor with the metric. Note that in general, the expression for the components of the Riemann curvature tensor at a point involves not only the components of the metric at that point, but also their first and second derivatives.

Recall, next, the notion of the sectional curvature of (M, g). Let $p \in M$ and let π be a plane in $T_p M$ (the tangent space at p) that is either spacelike (meaning that the restriction of g to π has signature $(+, +)$) or timelike (meaning that the restriction of g to π has signature $(-, +)$).[3] Let $v, w \in T_p M$ form an

---

[2] For helpful treatments of some aspects of general relativity in terms of sectional curvature, see Santander, Nieto, and Cordero, "A Curvature-Based Derivation of the Schwarzschild Spacetime"; and Mumford, "Ruminations on Cosmology and Time," §3.

[3] Here we extend the usual Lorentzian terminology to cover the Riemannian case: if g is Riemannian, then all tangent planes are spacelike; when g is a Lorentz metric, some tangent

orthonormal basis for π. The *sectional curvature* of π is defined as:

$$k(p,\pi) := \delta \cdot R_{abcd} w^a v^b w^c v^d,$$

where $\delta = 1$ if π is spacelike and $\delta = -1$ if π is timelike (the value of $k(p,\pi)$ is independent of the choice of orthonormal basis).[4] Note that if the Riemann curvature tensor of a manifold vanishes, then so do all of its sectional curvatures.

In the case where π is spacelike, its sectional curvature can be characterized alternatively as follows. For v a unit vector in π, let $x(v,r)$ be the result of tracing r units of distance along the geodesic through p with tangent vector v. For r sufficiently small, as v varies over unit-length vectors in π, $x(v,r)$ traces out a closed curve in M. Let $C(r)$ denote the length of this curve. We expect that for very small r, $C(r)$ will be approximately $2\pi r$. The sectional curvature $k(p,\pi)$ measures lowest-order (=third-order) disparity between $C(r)$ and $2\pi r$ as $r \to 0$.[5]

Let (M,g) be a Riemannian or Lorentz manifold, p be a point in M, and π a non-degenerate tangent plane in $T_p M$. For sufficiently small $\varepsilon > 0$, the geodesics of length ε that depart from p with a direction in π fill a two-dimensional submanifold U of M. We can ask how initially parallel nearby geodesics in U near p behave—do they converge, remain parallel, or diverge?[6] The answer depends on the sign of k. If π is spacelike, then initially parallel nearby geodesics in U near p diverge if $k(\pi,p)$ is negative, remain parallel if $k(\pi,p)$ vanishes, and converge if $k(\pi,p)$ is positive. If π is timelike, then initially parallel nearby timelike geodesics in U near p converge if $k(\pi,p)$ is negative, remain parallel if $k(\pi,p)$ vanishes, and diverge if $k(\pi,p)$ is positive.

A Riemannian or Lorentz manifold (M,g) is called a *space of constant curvature* if $k = k(p,\pi)$ is independent of both p and π.[7] In this case, the Riemann curvature tensor assumes the following simple form:

---

planes are spacelike, some are timelike, and some neither (the null case in which the restriction of g yields a degenerate bilinear form).

[4] For the formula for an arbitrary basis, see, e.g., Beem, Ehrlich, and Easley, *Global Lorentzian Geometry*, equation 2.37.

[5] See, e.g., Bishop and Goldberg, *Tensor Analysis on Manifolds*, §5.14.

[6] For details and precision, see, e.g., Beem, Ehrlich, and Easley, *Global Lorentzian Geometry*, §2.4.

[7] When dim $M \geq 3$, if $k(p,\pi)$ is independent of π at each $p \in M$ then it is independent of both p and π. See, e.g., Wolf, *Spaces of Constant Curvature*, corollary 2.2.7.

$$R_{abcd}(x) = k[g_{bc}(x)g_{ad}(x) - g_{ac}(x)g_{bd}(x)].$$

Note the contrast with the general case: in a space of constant curvature, the components of the Riemann tensor at a point is determined by the components of the metric at that point (without requiring recourse to their derivatives). Note also that if all the sectional curvatures of a manifold vanish, then so does its Riemann curvature tensor.

It is a fundamental fact that for each $n \geq 2$ and each $k \in \mathbb{R}$, each n-dimensional Riemannian space of constant curvature k has the same local geometry, and likewise for Lorentz spaces of constant curvature.[8]

So, for example, each flat ($k = 0$) Riemannian two manifold looks locally like the Euclidean plane. Of course, there is a large variety of flat surfaces. Only a few of these are geodesically complete (alongside the Euclidean plane itself, various tori, Klein bottles, cylinders, and Möbius bands). There are many incomplete examples—the interior of any blob in the plane is a flat surface in its own right, and so is the region exterior to the blob. An instructive example is the cone: cut an infinite wedge-shaped region out of the plane and identify the edges of the remaining region—the result will be a flat surface, with a singularity at its central point.[9] Each of these has the same local geometry—so we might as well say that they all have the same local geometry as a Klein bottle or of a cone, as say that they all have the same local geometry as the Euclidean plane.

REMARK 3.1 (Sectional Curvature of Product Spaces). Let (M_1, g_1) be a Lorentz manifold or a Riemannian manifold and let (M_2, g_2) be a Riemannian manifold.[10] Then the *product* of (M_1, g_1) and (M_2, g_2) is the manifold $M = M_1 \times M_2$ equipped with the metric $g = g_1 \oplus g_2$.[11] If a Lorentz or Riemannian manifold (M, g) is a product of some such (M_1, g_1) and (M_2, g_2)

---

[8] See, e.g., Wolf, *Spaces of Constant Curvature*, theorem 2.4.11.

[9] We can do the same sort of thing in the Lorentz case. Beginning with Minkowski spacetime, fix an inertial frame and delete the same infinite wedge-shaped region from each surface of constant time, then identify points along the boundary in the natural way. The resulting flat spacetime has a timelike singularity (corresponding to points at the spatial origin of the chosen inertial frame) and provides a basic model of cosmic strings—see, e.g., Griffiths and Podolský, *Exact Space-Times in Einstein's General Relativity*, §3.4.

[10] We allow the case dim $M_i = 1$. Each one-dimensional manifold is topologically either a line or a circle. In one dimension, Lorentz metrics are negative-definite.

[11] That is: g is the metric on M such that for any $p = (p_1, p_2)$ in M and any (v_1, v_2) and (w_1, w_2) in $T_p M = T_{p_1} M_1 \oplus T_{p_2} M_2$,

$$g(p_1, p_2)[(v_1, v_2), (w_1, w_2)] = g_1(p_1)[(v_1, w_1)] + g_2(p_2)[(v_2, w_2)].$$

then some of its sectional curvatures will vanish: whenever $v \in T_{p_1} M_1$ and $w \in T_{p_2} M_2$, the sectional curvature of the tangent plane spanned by $(v, 0)$ and $(0, w)$ at $(p_1, p_2) \in M$ is zero (if defined).[12] So, in particular, the only way a product space can have constant curvature is if it is flat (a fact that it will be handy to keep in mind below).

EXERCISE 3.1 (Easier). The metric tensor g of a Riemannian or Lorentz manifold determines the Riemann tensor and the sectional curvature of each tangent plane. It is natural to wonder whether the converse entailments hold.

a) Let M be the unit sphere in Euclidean space. Let g_1 be the Riemannian metric on M induced by the ambient Euclidean metric. Let $g_2 := 2g_1$. Show that g_1 and g_2 have the same Riemann curvature tensor but do not assign the same sectional curvature to any tangent plane.

b) Let M be any manifold, let g be a Riemannian or Lorentz metric on M, and let $\lambda > 1$. Show that g and $\lambda \cdot g$ have the same Riemann curvature tensor.

c) It follows from part (a) that knowing the Riemann curvature tensor of a Riemannian manifold does not suffice to determine its sectional curvature facts. How is this consistent with the our definition of sectional curvature?

d) Let M be a three-dimensional torus. Show that there are Riemannian metrics g_1 and g_2 such that: (i) the two metrics agree about all sectional curvature facts; (ii) (M, g_1) and (M, g_2) are not isometric.[13]

(Moral: in some cases, at least, complete knowledge of sectional curvature or Riemann curvature fails to determine the metric tensor up to isometry. But setting aside the two-dimensional case, in both the Riemannian setting and the Lorentzian setting: generically, Riemann curvature determines the metric up to isometry and a scale factor; generically, sectional curvature determines the metric up to isometry.[14])

---

[12] See, e.g., O'Neill, *Semi-Riemannian Geometry*, corollary 3.58.
[13] HINT: make life easy—make both metrics to be flat. How then might (M, g_1) and (M, g_2) differ?
[14] See Berger, *A Panoramic View of Riemannian Geometry*, §4.5; and Hall, *Symmetries and Curvature Structure in General Relativity*, chapter 9.

3. Highly Symmetric Spaces

In this section, M is an arbitrary n-dimensional Riemannian or Lorentz manifold (with $n \geq 2$). We denote by $I(M)$ the group of isometries of M. Note that $I(M)$ has a natural Lie structure (i.e., it is both a group and a manifold, with the two structures meshing).[15]

So, in this context, the size (= dimension) of the isometry group of a space is a well-defined. And, in fact, we have an upper bound on this size: $\dim I(M) \leq n(n+1)/2$.[16] Here is a rough intuitive argument for this claim. Isometries are 'rigid' transformations: in particular, for any point $x \in M$, if two isometries of (M, g) agree about which point $y \in M$ they map x to and about how their differentials act on tangent vectors at x, then they must in fact coincide.[17] So even if isometries were as plentiful as they could possibly be, we could always determine one by first specifying a point $y \in M$ and then specifying an orthogonal transformation from $T_x M$ to $T_y M$. So

$$\dim I(M) \leq n + \frac{1}{2} n(n-1),$$

since (i) $\dim M = n$ and (ii) the orthogonal group of for an n-dimensional space has dimension $\frac{1}{2} n(n-1)$.[18]

We introduce a trio of global symmetry conditions.

MAXIMAL SYMMETRY. M is *maximally symmetric* if $\dim\ I(M) = n(n+1)/2$.

ISOTROPY. M is *isotropic* if at at each $x \in M$, for any two tangent vectors of the same non-zero length at x, there is an isometry of M that fixes x and whose differential maps the first tangent vector to the second.

HOMOGENEITY. M is *homogeneous* if for any two points of M, there is an isometry that maps the first to the second.

---

[15] See, e.g., Kobayashi, *Transformation Groups in Differential Geometry*, theorem I.3.1.
[16] See, e.g., O'Neill, *Semi-Riemannian Geometry*, lemma 9.28 and proposition 9.33; or Kobayashi, *Transformation Groups*, theorem II.3.1.
[17] O'Neill, *Semi-Riemannian Geometry*, proposition 3.62.
[18] For (ii), see, e.g., O'Neill, *Semi-Riemannian Geometry*, 235.

Each of these symmetry conditions also comes in a local variant.

LOCAL MAXIMAL SYMMETRY. M is *locally maximally symmetric* if it admits $n(n+1)/2$ Killing vector fields.[19]

LOCAL ISOTROPY. M is *locally isotropic* if for $x \in M$ and any two tangent vectors of the same non-zero length at x, there is a neighbourhood U of x and an isometry of U that fixes x and whose differential maps the first tangent vector to the second.

LOCAL HOMOGENEITY. M is *locally homogeneous* if for any two points $x, y \in M$, there are neighbourhoods U and V of x and y and an isometry from U to V that maps x to y.

The charts below summarize the relations of implication between these conditions (along with the condition of having constant curvature) for an n-dimensional Lorentz or Riemannian manifold. A bi-directional arrow indicates that two properties are equivalent for any $n \geq 2$. A uni-directional arrow pointing from one condition to another indicates that the first implies the second for any $n \geq 2$; if the arrow is not marked by an asterisk or a dagger, then the converse fails for each $n \geq 2$; an asterisk indicates that the converse implication holds for $n = 2$, but not otherwise; a dagger indicates that the converse implication holds for $n = 2$ and for n odd, but not otherwise.

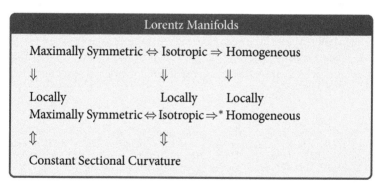

---

[19] This is again the maximum number possible. See, e.g., O'Neill, *Semi-Riemannian Geometry*, lemma 9.28.

Riemannian Manifolds

Maximally Symmetric $\Rightarrow^\dagger$ Isotropic $\Rightarrow$ Homogeneous

$\Downarrow$ $\qquad\qquad\qquad$ $\Downarrow$ $\qquad$ $\Downarrow$

Locally $\qquad\qquad\qquad$ Locally $\qquad$ Locally
Maximally Symmetric $\Rightarrow^\dagger$ Isotropic $\Rightarrow^*$ Homogeneous

$\Updownarrow$ $\qquad\qquad\qquad\qquad$ $\Uparrow^\dagger$

Constant Sectional Curvature

For the diligent, here are some facts, references, and exercises to back up (and extend) the charts. Let us begin with what is common to the Riemannian case and the Lorentz case.

1. Maximal symmetry implies isotropy: Compare the classification theorems for maximally symmetric Riemannian and Lorentz manifolds with the classification theorem for isotropic Riemannian and Lorentz manifolds.[20]
2. Isotropy implies homogeneity.[21] But the converse doesn't hold—see Exercise 3.2 below.[22]
3. Each of maximal symmetry, isotropy, and homogeneity implies the corresponding local condition (immediate from the definition) but none of the converse implications hold (see Exercise 3.3 below).
4. Local maximal symmetry is equivalent to being a space of constant curvature.[23]
5. Having constant curvature implies being locally isotropic: we will see in Sections 4 and 5 below that every Riemannian or Lorentz manifold of

---

[20] For the maximally symmetric Riemannian manifolds, see Kobayashi, *Transformation Groups*, theorem II.3.1. For the maximally symmetric Lorentz manifolds, see Patrangenaru, "Lorentz Manifolds with the Three Largest Degrees of Symmetry," theorem 3.1. For the isotropic manifolds of both signatures, see Wolf, *Spaces of Constant Curvature*, theorem 12.4.5.

[21] See, e.g., Wolf, *Spaces of Constant Curvature*, lemma 11.6.6.

[22] In an isotropic Riemannian manifold, any direction at one point is geometrically equivalent to any direction at any other point—so for any two geodesics, there is an isometry that maps one to the other. In the Lorentz case, this becomes: each timelike geodesic is geometrically equivalent to each other timelike geodesic, and likewise for spacelike geodesics. For discussion, see Beem, "Lorentzian Distance and Curvature," §3. In fact, for Lorentz manifolds timelike isotropy (the equivalence between any two unit timelike vectors at each point) is equivalent to isotropy and local timelike isotropy is equivalent to local isotropy—see Tits, "Le principe d'inertie en relativité générale," théorème 1′ and proposition 4.

[23] See Eisenhart, *Riemannian Geometry*, §27.

SYMMETRY AND CURVATURE 51

constant curvature shares its local geometry with an isotropic space—
so every space of constant curvature is locally isotropic.

6. Local maximal symmetry implies local isotropy: by the preceding two points.

7. Local isotropy implies local homogeneity.[24] What about the converse?

   a) When $n = 2$, local homogeneity implies local isotropy (Exercise 3.4).

   b) But for $n > 2$, there are Riemannian and Lorentz n-manifolds that locally homogeneous without being locally isotropic. Here is one way to see this.

      i. The rank-1 symmetric manifolds are Riemannian manifolds that can be viewed as relatives of hyperbolic and elliptic geometries and which occur in each even dimension greater than two.[25] Each rank-1 symmetric space is isotropic. Each has either everywhere positive sectional curvature or everywhere negative sectional curvature, but none has constant curvature.[26]

      ii. Every locally isotropic Riemannian or Lorentz manifold is a space of constant curvature or has the same local geometry as a rank-1 symmetric space.[27]

      iii. So no locally isotropic Lorentz or Riemannian manifold has vanishing sectional curvature in some tangent plane and non-vanishing sectional curvature in some other tangent plane. But for each $n \geq 3$, there are locally homogeneous Lorentz n-manifolds and locally homogeneous Riemannian manifolds with this feature (Exercise 3.5).

Consider, next, the aspects where the Lorentzian and Riemannian pictures differ.

A) Does isotropy imply maximal symmetry?

   * Yes, for Lorentz manifolds of any dimension $n \geq 2$: compare the classification theorem for maximally symmetric Lorentz manifolds with the classification theorem for isotropic Lorentz manifolds.[28]

---

[24] See, e.g., Wolf, *Spaces of Constant Curvature*, theorem 12.3.1.(i).

[25] For an overview, see Berger, *Panoramic View*, §4.3.5.

[26] See, e.g., Petersen, *Riemannian Geometry*, chapter 10. The most familiar examples of such space are probably the complex projective spaces that arise as the spaces of states for multi-qubit systems—see, e.g., Brody and Hughston, "Geometric Quantum Mechanics."

[27] See Wolf, *Spaces of Constant Curvature*, Theorem 12.3.1.

[28] See theorem 3.1 of Patrangenaru, "Lorentz Manifolds"; and theorem 12.4.5 of Wolf, *Spaces of Constant Curvature*.

* Yes, for Riemannian n-manifolds with $n = 2$ or odd: compare the classification theorem for maximally symmetric Riemannian manifolds with the classification theorem for isotropic Riemannian manifolds.[29]
* No, for Riemannian manifolds in even dimension four and higher: we know from 7(b)i above that in these dimensions, there are rank-1 symmetric manifolds that are isotropic without having constant curvature—and which therefore cannot be maximally symmetric.

B) Does local isotropy imply constant curvature? Does local isotropy imply local maximal symmetry?
* Yes, for Lorentz manifolds of dimension $n \geq 2$ and for Riemannian n-manifolds with $n = 2$ or odd: in these cases, all locally isotropic spaces are spaces of constant curvature (and hence also locally maximally symmetric).[30]
* No, for Riemannian n-manifolds with $n > 2$ and even: as noted in 7(b)i above, in these dimensions there exist rank-1 symmetric Riemannian manifolds that are isotropic (and so locally isotropic) but which do not have constant curvature (and so are not locally maximally symmetric either).

EXERCISE 3.2 (Easier). Homogeneity does not imply isotropy.
a) Given an example of a homogeneous but not isotropic Riemannian manifold.
b) Given an example of a homogeneous but not isotropic Lorentz manifold.
c) Give examples of both kinds in each dimension $n \geq 2$.

EXERCISE 3.3 (Easier). Show that none of our trio of local symmetry conditions implies any of our trio of global symmetry conditions.[31]

EXERCISE 3.4 (Medium). Let M be a locally homogeneous two-dimensional Riemannian or Lorentz manifold.
a) Show that M is a space of constant curvature.
b) Conclude that M is locally isotropic.

---

[29] See theorem II.3.1 of Kobayashi, *Transformation Groups*; and theorem 12.4.5 of Wolf, *Spaces of Constant Curvature*.

[30] See theorem 12.3.1 of Wolf, *Spaces of Constant Curvature*.

[31] HINT: for each $n \geq 2$, find examples of locally maximally symmetric Lorentz and Riemannian manifolds that are not homogeneous.

EXERCISE 3.5 (Medium). Call a Lorentz or Riemannian manifold *ragged* if some of its tangent planes have vanishing sectional curvature while others have non-vanishing sectional curvature.

    a) Show that the Einstein static universe in four dimensions is ragged.[32]

    b) Construct locally homogeneous but ragged Lorentz and Riemannian n-manifolds for each $n \geq 3$.

EXERCISE 3.6 (Frame-Homogeneity and Strong Isotropy.). We note here two other conditions that figure in the literature. Let M be a Riemannian or Lorentz manifold.

FRAME-HOMOGENEITY. M is *frame-homogeneous* if for any two points in M and any two orthonormal frames at those points, there is an isometry of M that maps the first point to the second and whose differential maps the first frame to the second.

STRONG ISOTOPY. M is *strongly isotropic* if at each point of M, for any two orthonormal frames at that point, there is an isometry of M that fixes the given point and whose differential maps the first frame to the second.

Prove that these are equivalent to maximal symmetry by showing the following.

    a) (Medium). Strong isotropy implies frame-homogeneity.[33]

    b) (Medium). Frame-homogeneity implies maximal symmetry.[34]

    c) (Harder). Maximal symmetry implies strong isotropy.[35]

4. The Riemannian Case

Recall that for each $n \geq 2$ and each $k \in \mathbb{R}$, any two n-dimensional Riemannian manifolds of constant curvature k share their local geometry. We here introduce a paradigm example for each value of n and k.

---

[32] HINT: recall Remark 3.1 above.
[33] HINT: observe, first, that strong isotropy implies homogeneity.
[34] HINT: try a dimension-counting argument.
[35] HINT: consult the classification theorems cited in fn. 20 above.

These paradigms admit simple characterizations as subsets of scalar product spaces, which makes their symmetries easy to characterize.[36] For any $n \geq 2$ and any $0 \leq \nu \leq n$, we denote by $\mathbb{R}^n_\nu$ the set of n-tuples of real numbers equipped with the scalar product:

$$\langle (x_0, \ldots, x_{n-1}), (y_0, \ldots, y_{n-1}) \rangle = -\sum_{i=0}^{\nu-1} x_i y_i + \sum_{j=\nu}^{n-1} x_j y_j.$$

A scalar product can be thought of as a not-necessarily-positive-definite inner product. Note that $\mathbb{R}^n_0$ is just $\mathbb{R}^n$ equipped with its standard inner product. And $\mathbb{R}^n_1$ is n-dimensional Minkowski spacetime. By extension, for any $\mathbb{R}^n_\nu$, we call $x \in \mathbb{R}^n_\nu$ *timelike* if $\langle x, x \rangle < 0$ and *spacelike* if $\langle x, x \rangle > 0$. We denote by $O_\nu(n)$ the (semi-orthogonal) group of transformations that preserves the scalar product of $\mathbb{R}^n_\nu$.[37] Each $O_\nu(n)$ is a Lie group of dimension $n(n-1)/2$. When $\nu = 0$ or $\nu = n$, all vectors are spacelike or all vectors are timelike and $O_\nu(n)$ has two components (each transformation either does or doesn't preserve orientation). When $0 < \nu < n$, $\mathbb{R}^n_\nu$ features both timelike and spacelike vectors and $O_\nu(n)$ has four components (each transformation does or doesn't preserve orientation in maximal subspaces spanned by timelike vectors and does or doesn't preserve orientation in maximal subspaces spanned by spacelike vectors).

For each $n \geq 2$ and each $k \in \mathbb{R}$ we have the following paradigms of Riemannian spaces of constant curvature (classified by the sign of k):

SPHERICAL SPACES ($k > 0$). The *spherical space* $S_n(r)$ of dimension n and radius $r > 0$ is the set

$$\{x \in \mathbb{R}^{n+1}_0 \mid \langle x, x \rangle = r^2\},$$

equipped with the Riemannian metric induced by the scalar product of the ambient $\mathbb{R}^{n+1}_0$. $S_n(r)$ is a space of constant curvature $k = 1/r^2$.[38] The isometry group of $S_n(r)$ is $O_0(n+1)$ (the group of rotations and reflections that

---

[36] On scalar product spaces, see, e.g., O'Neill, *Semi-Riemannian Geometry*, 47 ff.

[37] On semi-orthogonal groups, see, e.g., O'Neill, *Semi-Riemannian Geometry*, 233 ff.

[38] See, e.g., O'Neill, *Semi-Riemannian Geometry*, proposition 4.29(1).

preserve the origin in the ambient Euclidean space), a group of $n(n+1)/2$ dimensions.[39]

EUCLIDEAN SPACES ($k = 0$). The n-dimensional Euclidean space, $\mathbb{R}_0^n$, which we denote $\mathbb{E}_n$. Each $\mathbb{E}_n$ is flat (i.e., has vanishing sectional curvature). The isometry group of $\mathbb{E}_n$ is the (semi-direct) product of $O_0(n)$ (rotations and reflections) with $\mathbb{R}^n$ (translations), a group of $n(n+1)/2$ dimensions.[40]

HYPERBOLIC SPACES ($k < 0$). The n-dimensional *hyperbolic space* $\mathbb{H}_n(r)$ of radius $r > 0$ is the set

$$\{x \in \mathbb{R}_1^{n+1} \,|\, \langle x, x \rangle = -r^2, x_0 > 0\},$$

equipped with the Riemannian metric induced by the scalar product of the ambient $\mathbb{R}_1^{n+1}$.[41] Thinking of $\mathbb{R}_1^{n+1}$ as $(n+1)$-dimensional Minkowski spacetime, $\mathbb{H}_n(r)$ is the set of points lying $1/r$ units of proper time to the future of the origin. $\mathbb{H}_n(r)$ is a space of constant curvature $k = -1/r^2$.[42] The isometry group of $\mathbb{H}_n(r)$ is that portion of $O_1(n+1)$ (the Lorentz group of the ambient Minkowski space) that preserves temporal orientation and so fixes $\mathbb{H}_n(r)$ as a set.[43] This is a group of $n(n+1)/2$ dimensions.

Scads of spaces of constant curvature can be manufactured by beginning with one of our paradigms and applying standard constructions such as making excisions and identifications. But typical spaces of constant curvature differ from the paradigms in admitting few isometries.

In each dimension $n \geq 2$, the hyperbolic spaces $\mathbb{H}_n(r)$ are the only homogeneous spaces of constant negative curvature—and hence also the only isotropic spaces of constant negative curvature and the only maximally symmetric spaces of this kind.[44]

---

[39] See, e.g., O'Neill, *Semi-Riemannian Geometry*, proposition 9.8.
[40] See, e.g., O'Neill, *Semi-Riemannian Geometry*, proposition 9.10.
[41] This model of hyperbolic geometry played an important role already in the nineteenth century—so as a geometrical object lacking physical interpretation, Minkowski spacetime was in the geometer's toolbox well before the advent of the special theory of relativity. See, e.g., Killing, *Die nicht-euklidischen Raumformen in analytischer Behandlung*, §§80 ff.; for further references and discussion, see Ratcliffe, *Foundations of Hyperbolic Geometry*, §3.6.
[42] See, e.g., O'Neill, *Semi-Riemannian Geometry*, proposition 4.29(2).
[43] See, e.g., O'Neill, *Semi-Riemannian Geometry*, corollary 9.9.
[44] See Wolf, *Spaces of Constant Curvature*, theorem 2.7.1.

The situation is a little more nuanced in the flat case ($k = 0$). In any dimension $n \geq 2$, the flat homogeneous spaces are Euclidean space and various n-dimensional versions of cylinders and tori: an n-dimensional flat homogeneous space has topology $\mathbb{R}^\ell \times \mathbb{T}^m$, where $\ell + m = n$ and $\mathbb{T}^m$ is the m-dimensional torus.[45] But for any $n \geq 2$, the $\mathbb{E}_n$ are the only isotropic flat n-dimensional spaces (and hence also the only maximally symmetric ones).[46]

In the $k > 0$ case, things are different yet again. The spherical spaces have highly-symmetric rivals.

ELLIPTIC SPACES. The *elliptic space* $\bar{S}_n(r)$ of dimension n and of radius $r > 0$, is the Riemannian space that results from identifying antipodal points in $S_n(r)$.[47] Each $\bar{S}_n(r)$ shares its local geometry with the corresponding $S_n(r)$ and so, in particular, is a space of constant curvature $k = 1/r^2$.[48] But the two spaces have different topologies: $S_n(r)$ is simply connected, $\bar{S}_n(r)$ is not.[49] Each $\bar{S}_n(r)$ is maximally symmetric (and hence also isotropic and homogeneous).[50]

In each dimension, the spherical and elliptic spaces are the only isotropic spaces of positive constant curvature (and hence also the only maximally symmetric spaces of positive constant curvature).[51] When the dimension

[45] See, e.g., Wolf, *Spaces of Constant Curvature*, theorem 2.7.1.
[46] See, e.g., Wolf, *Spaces of Constant Curvature*, corollary 8.12.9.
[47] That is: thinking of $S_n(r)$ as a subset of the $\mathbb{R}^{n+1}$, identify points in $S_n(r)$ related by the antipodal map $-I : (x_0, x_1, \ldots, x_n) \mapsto (-x_0, -x_1, \ldots, -x_n)$.
[48] On the development of elliptic geometry in the nineteenth century, see, e.g., Hawkins, *Emergence of the Theory of Lie Groups*, chapter 4.
[49] Also: $\bar{S}_n(r)$ is orientable if and only n is even. See Remark 4.1 below.
[50] Here is a fundamental result we will appeal to a number of times:

> Let M be a simply connected Riemannian or Lorentz manifold. Let Γ be a countable group of isometries acting freely and properly on M and let $N(\Gamma)$ be the normalizer subgroup of Γ in the isometry group of M. Then the isometry group of the quotient manifold M/Γ is isomorphic to $N(\Gamma)/\Gamma$.

(Up to slight differences of terminology, this is proposition 9.20 of O'Neill, *Semi-Riemannian Geometry*. For the notion of free and proper actions, see Remark 8.1 below.)

In the present case, since the isometry group $O_0(n+1)$ of $S_n(r)$ consists of linear maps, each of them commutes with the antipodal map of fn. 47 (which is, after all, just scalar multiplication). So the normalizer in $O_0(n+1)$ of the two-element group generated by the antipodal map is all of $O_0(n+1)$. So the isometry group of $\bar{S}_n(r)$ is isomorphic to $O_0(n+1)/\mathbb{Z}_2$, which has the same dimension as $O_0(n+1)$.
[51] See Wolf, *Spaces of Constant Curvature*, theorem 8.12.9.

SYMMETRY AND CURVATURE 57

n is even, these are also the only homogeneous spaces of positive constant curvature—but in odd dimensions, there are further such spaces.[52]

So the only maximally symmetric Riemannian manifolds are the hyperbolic, Euclidean, spherical, and elliptic spaces.[53] Further, if $n = 2$ or n is odd, every isotropic Riemannian n-manifold must be one these classical geometries.[54]

EXERCISE 3.7 (Easier). Let R be a rectangular region of $\mathbb{R}^2$ (equipped with the standard Euclidean metric) with sides of length a and b ($a < b$).

  a) We can construct a torus $\mathbb{T}$ by identifying points along opposite edges of R (identifying points along the vertical edges if and only if they lie on the same horizontal line and identifying points along the horizontal edges if and only if they lies on the same vertical line). Argue that $\mathbb{T}^2$ is homogeneous but not isotropic.

  b) We can construct a Klein bottle $\mathbb{K}$ by identifying points along opposite edges of R, giving a twist to the horizontal edges (identifying points along the vertical edges if and only if they lie on the same horizontal line; identify a point on the top edge lying a units of distance from the top-left corner with the point on the bottom edge lying a units of distance from the bottom-right corner). Argue that $\mathbb{K}$ is neither homogeneous nor isotropic.

  c) Do your arguments work if $a = b$ (so that R is a square)?

5. The Lorentz Case

As in the Riemannian case, for any $n \geq 2$ and k, any two n-dimensional Lorentz manifolds of constant curvature k share their local geometries. For each such n and k, we introduce a paradigm n-dimensional Lorentz manifold of constant curvature k. Each of these paradigms is maximally symmetric (and so also isotropic and homogeneous). As in

---

[52] For the classification of homogeneous spaces of constant positive curvature, see Wolf, *Spaces of Constant Curvature*, theorem 2.7.1.

[53] Recall from above that any maximally symmetric Riemannian manifold must have constant curvature.

[54] Recall that for such n, isotropy is equivalent to maximal symmetry.

Section 4 above, our paradigms again arise as simple subsets of scalar product spaces.[55]

DE SITTER SPACETIMES ($k > 0$). The *de Sitter spacetime* $dS_n(r)$ of dimension n and radius $r > 0$ is the set

$$\{x \in \mathbb{R}_1^{n+1} \,|\, \langle x, x \rangle = r^2\},$$

equipped with the Lorentz metric induced by the scalar product of the ambient $\mathbb{R}_1^{n+1}$. Each $dS_n(r)$ is a space of constant curvature $k = 1/r^2$ and a solution of the vacuum Einstein equations with cosmological constant $\Lambda = n(n-1)/2r^2$.[56] The isometry group of $dS_n(r)$ is $O_1(n+1)$, the Lorentz group of $(n+1)$-dimensional Minkowski spacetime.[57] This is a group of $n(n+1)/2$ dimensions.

MINKOWSKI SPACETIMES ($k = 0$). The Minkowski spacetime $\mathbb{M}_n$ in n dimensions is $\mathbb{R}_1^n$. Each $\mathbb{M}_n$ is a flat solution of the $\Lambda = 0$ vacuum Einstein equations. The isometry group of $\mathbb{M}_n$ is the Poincaré group, the (semi-direct) product of the Lorentz group $O_1(n)$ with the translation group.[58] This is a group of $n(n+1)/2$ dimensions.

ANTI-DE SITTER HYPERBOLOIDS ($k < 0$). The *anti-de Sitter hyperboloid* $\underline{AdS}_n(r)$ of dimension n and radius $r > 0$ is the set

$$\{x \in \mathbb{R}_2^{n+1} \,|\, \langle x, x \rangle = -r^2\},$$

equipped with the Lorentz metric induced by the scalar product of the ambient $\mathbb{R}_2^{n+1}$. Each $\underline{AdS}_n(r)$ is a space of constant curvature $k = -1/r^2$ and a solution of the vacuum Einstein equations with $\Lambda = -n(n-1)/2r^2$.[59] The isometry group of $\underline{AdS}_n(r)$ is $O_2(n+1)$.[60] This is a group of $n(n+1)/2$ dimensions.

Note a few points of similarity and difference between the de Sitter case and the anti-de Sitter case (see Figure 3.1). In direct analogy with the de Sitter

---

[55] Note that the relation between the characterizations of the Riemannian paradigms and those of the Lorentz paradigms—to pass from the former to the latter you augment ν by one.
[56] See, e.g., O'Neill, *Semi-Riemannian Geometry*, proposition 4.29(1).
[57] See, e.g., O'Neill, *Semi-Riemannian Geometry*, proposition 9.8.
[58] See, e.g., O'Neill, *Semi-Riemannian Geometry*, proposition 9.10.
[59] See, e.g., O'Neill, *Semi-Riemannian Geometry*, proposition 4.29(2).
[60] See, e.g., O'Neill, *Semi-Riemannian Geometry*, proposition 9.8.

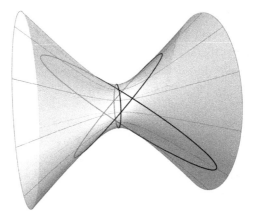

Figure 3.1 The anti-de Sitter hyperboloid in two spacetime dimensions with three closed timelike geodesics highlighted.

case: anti-de Sitter geodesics arises as the intersection of the anti-de Sitter hyperboloid with planes through the origin of the ambient $\mathbb{R}_2^{n+1}$.[61] But for each $n \geq 2$, whereas each $dS_n(r)$ has topology $\mathbb{R} \times S^{n-1}$, each $\underline{AdS}_n(r)$ has topology $S^1 \times \mathbb{R}^{n-1}$.[62] And while each de Sitter spacetime is of course globally hyperbolic (with the $\mathbb{R}$ factor in $\mathbb{R} \times S^{n-1}$ corresponding to time), each anti-de Sitter hyperboloid features closed timelike geodesics (with the S^1 factor in $S^1 \times \mathbb{R}^{n-1}$ corresponding to time).

Note that when $n = 2$, de Sitter spacetimes and the anti-de Sitter hyperboloids have the same topology (that of a cylinder)—and, indeed, the metrics of $dS_2(r)$ and $\underline{AdS}_2(r)$ are the same up to the transformation that switches the roles of space and time so that in two spacetime dimensions de Sitter spacetimes and anti-de Sitter hyperboloids are essentially the same object.[63]

---

[61] See the treatment of O'Neill, *Semi-Riemannian Geometry*, 112 f.

[62] See, e.g., O'Neill, *Semi-Riemannian Geometry*, lemma 4.24.

[63] As a geometric object without physical interpretation, this appears already in an 1887 paper of Poincaré, "Sur les hypothèses fondamentales de la géométrie"; for further discussion, see Sternberg, "Review of: *Imagery in Scientific Thought* by Arthur I Miller." For an early and comprehensive mathematical treatment of the geometry of hyper-quadrics (such as de Sitter and anti-de Sitter hyperboloids) as spaces of constant curvature, see Eisenhart, *Riemannian Geometry*, §§61 ff.

In the literature on general relativity, the earliest discussion of spacetimes of constant negative curvature appears to be in Silberstein, "General Relativity without the Equivalence Hypothesis." However, Silberstein does not manage to find the anti-de Sitter metric (see Exercise 5.1).

In showing how the Dirac equation can be formulated in de Sitter spacetime, Dirac casually mentions that his approach carries over to the $k < 0$ setting—see "Electron Wave Equation in de-Sitter Space," §2. I do not know whether this is the earliest discussion of anti-de Sitter

Anti-de Sitter hyperboloids, Minkowski spacetimes, and de Sitter space-times are Lorentz manifolds of maximal symmetry. If the preceding discussion has done its job, the reader suspects that they are not the only ones. The reader is right. (In what follows it is sometimes convenient to restrict attention to dimension $n \geq 3$—the two-dimensional case is addressed in Remark 3.2.)

In the $k = 0$ regime, the picture is much as in the Riemannian case. In each dimension $n \geq 2$, Minkowski spacetime is the only maximally symmetric flat Lorentz manifold (and hence is also the only isotropic flat Lorentz manifold).[64] There are, however, further homogeneous flat Lorentz manifolds. In parallel with the Riemannian case, each complete and homogeneous flat Lorentz manifold arises as a quotient of Minkowski spacetime by a discrete group of translations.[65]

When when $k > 0$, we again find something similar to the Riemannian case. In each scalar product space $\mathbb{R}^n_\nu$ the *antipodal map* is the map $-I : (x_0, \ldots, x_{n-1}) \mapsto (-x_0, \ldots, -x_{n-1})$. By analogy with our introduction of the elliptic space corresponding to any spherical space, we can also introduce elliptic analogs of de Sitter spacetimes:

ELLIPTIC DE SITTER SPACETIMES. The *elliptic de Sitter spacetime* $\overline{dS}_n(r)$ of dimension $n \geq 2$ and radius $r > 0$ is is the result of identifying antipodal points in $dS_n(r)$.[66] Each $\overline{dS}_n(r)$ shares its local geometry with $dS_n(r)$ and so in particular is a space of constant curvature $k = 1/r^2$ and a solution of the $\Lambda = (n-1)/r^2$ vacuum Einstein field equations. But the two spaces have different global structures: $dS_n(r)$ is time-orientable but $\overline{dS}_n(r)$ is not (see Section 4.3 below). Each $\overline{dS}_n(r)$ is maximally symmetric.[67]

physics. The earliest occurrence of the label *anti-de Sitter* known to me is in Calabi and Markus, "Relativistic Space Forms."

[64] For maximal symmetry, see theorem 3.1 of Patrangenaru, "Lorentz Manifolds." For isotropy, see theorem 11.6.8 of Wolf, *Spaces of Constant Curvature*.

[65] See Wolf, *Spaces of Constant Curvature*, theorem 3.7.11. Recall that although homogeneity implies completeness in the Riemannian setting, there are homogeneous but incomplete Lorentz manifolds—see Wolf, *Spaces of Constant Curvature*, §2.8. For a treatment of homogeneous but incomplete flat Lorentz manifolds, see Duncan and Ihrig, "Homogeneous Spaces of Zero Curvature."

[66] Klein appears to be the first to have explicitly noted the existence of elliptic de Sitter spacetime: see his letter to de Sitter of 19 April 1918 (quoted in de Sitter, "Further Remarks on the Solutions of the Field-Equations of Einstein's Theory of Gravitation," 1310), his postcard to Einstein of 25 April 1918 (8.518), and his letter to Einstein of 31 May 1918 (8.552). But see also the cryptic remarks of Schwarzschild in his letter to Einstein of 6 February 1916 (8.188).

[67] This follows via an argument parallel to the argument of fn. 50 above.

In dimension three and higher the only maximally symmetric (or isotropic) Lorentz manifolds of constant positive curvature are the de Sitter spacetimes and the elliptic Sitter spacetimes.[68] Indeed, in these dimensions, these are the only complete and homogeneous Lorentz spaces of constant positive curvature.[69]

The case of constant negative curvature is a little messier. Note, first, that given any $AdS_n(r)$, we can construct an infinite family of closely related Lorentz manifolds that share its local geometry: its double cover $AdS_n^{(2)}(r)$ (same topology, but with the period of time doubled); its triple cover $AdS_n^{(3)}(r)$ (same topology, period of time tripled); ...; its k-fold cover $AdS_n^{(k)}(r)$ (same topology, period of time increased k-fold); ...; and its universal cover $AdS_n(r)$ (topology of $\mathbb{R}^n$, time linear)—we will reserve the name *anti-de Sitter spacetime* for this last object. The construction will be discussed in more detail in Chapter 5 below. For now, it suffices to picture this family of objects as being analogous to a family of surfaces we could construct beginning from a cylinder of a given radius. We could take two copies of the cylinder, slit each along an edge parallel to the axis of symmetry to give us two infinite strips, then glue the edges of these strips to yield a cylinder of twice the radius of the given cylinder. Or we could begin with k copies of the given cylinder and perform these operations to construct a cylinder with k times the radius of the original cylinder. Or we could begin with a countably infinite family of copies of the cylinder, make a countably infinite family of strips, and glue them together to yield a copy of the Euclidean plane. For present purposes, the crucial point is that all of the covering spacetimes of a given anti-de Sitter hyperboloid are themselves maximally symmetric (and so also isotropic and homogeneous) Lorentz manifolds.[70] But wait—there is more.

ELLIPTIC ANTI-DE SITTER SPACETIMES. The *elliptic anti-de Sitter spacetime* $\overline{AdS}_n(r)$ of dimension $n \geq 3$ and radius $r > 0$ is is the result of identifying antipodal points in $AdS_n(r)$.[71] Each $\overline{AdS}_n(r)$ shares its local geometry with $AdS_n(r)$ and so in particular is a space of constant curvature $k = -1/r^2$ and a solution of the $\Lambda = -(n-1)/r^2$ vacuum Einstein field equations. But

---

[68] For maximal symmetry, see theorem 3.1 of Patrangenaru, "Lorentz Manifolds." For isotropy, see theorem 11.6.7 of Wolf, *Spaces of Constant Curvature*.

[69] See Wolf, *Spaces of Constant Curvature*, corollary 11.6.4.

[70] See fn. 4 in Chapter 5 below.

[71] See the discussion of the *Klein Model* in Bonsante and Seppi, "Anti-de Sitter Geometry and Teichmüller Theory."

the two spaces have different global structures.[72] Each $\overline{AdS}_n(r)$ is maximally symmetric.[73]

In each dimension $n \geq 2$, the only maximally symmetry (or isotropic) Lorentz manifolds of constant negative curvature are the $\overline{AdS}_n(r)$, $\underline{AdS}_n(r)$, $\underline{AdS}_n^{(k)}(r)$, and $AdS_n(r)$.[74] Each of these is homogeneous and complete.[75] But in each dimension $n \geq 3$ there exist further homogeneous and complete Lorentz spaces of constant negative curvature.[76]

The reader will be relieved to hear that we have now met all of the maximally symmetric Lorentz manifolds of dimension three or greater: de Sitter spacetimes and elliptic de Sitter spacetimes; Minkowski spacetime; anti-de Sitter hyperboloids, elliptic anti-de Sitter spacetimes, and the various coverings of the anti-de Sitter hyperboloids, including the anti-de Sitter spacetimes.[77] These are also all of the isotropic Lorentz spaces in each dimension $n \geq 3$.[78] These are all of course also homogeneous. But, as we have seen, they do not exhaust even the homogeneous Lorentz manifolds of constant curvature.[79] And there also exist physically interesting homogeneous Lorentz manifolds of non-constant curvature.[80]

REMARK 3.2 (The Two-Dimensional Case). Every connected manifold M has a universal covering space $\widetilde{M}$, which coincides with (=is diffeomorphic to) M if and only if M is simply connected.[81] An n-dimensional de Sitter spacetime has topology $\mathbb{R} \times S^{n-1}$ while an n-dimensional anti-de Sitter hyperbolid has topology $S^1 \times \mathbb{R}^{n-1}$. So anti-de Sitter hyperboloids are never simply connected and always admit non-trivial covers, while in dimension $n \geq 3$ de Sitter spacetimes are simply connected and do not admit non-trivial covers. But in two spacetime dimensions, de Sitter spacetimes, like anti-de

---

[72] See, e.g., Bonsante and Seppi, "Anti-de Sitter Geometry," §2.5.
[73] See fn. 4 in Chapter 5 below.
[74] For maximal symmetry, see theorem 3.1 of Patrangenaru, "Lorentz Manifolds." For isotropy, see theorem 11.6.7 of Wolf, *Spaces of Constant Curvature*.
[75] See Wolf, *Spaces of Constant Curvature*, lemma 11.6.6.
[76] See Wolf, *Spaces of Constant Curvature*, corollary 11.6.4.
[77] See theorem 3.1 of Patrangenaru, "Lorentz Manifolds."
[78] See Wolf, *Spaces of Constant Curvature*, theorem 12.4.5.
[79] See the references cited in fnn. 65 and 76.
[80] The Einstein static universe is one example. For further examples, see Stephani *et al.*, *Exact Solutions of Einstein's Field Equations*, chapter 12.
[81] For a review of the relevant concepts and results, see, e.g., O'Neill, *Semi-Riemannian Geometry*, appendix A; or Lee, *Introduction to Smooth Manifolds* (second edition). Springer (2013), 91 ff.

Sitter hyperboloids, have the topology of the cylinder. So any $dS_2(r)$ admits an infinite family of non-trivial covering spacetimes analogous to the universal cover of the anti-de Sitter hyperboloid and to its various k-fold covers.[82] Any such covering spacetime of a two-dimensional de Sitter spacetime is maximally symmetric (and hence also isotropic and homogeneous).[83] So in two spacetime dimensions, we need to add these covering spacetimes of de Sitter spacetime to the rosters of maximally symmetric and of isotropic Lorentz manifolds (and these are the only additions that need to be made).[84] In this dimension, the only homogeneous and complete Lorentz manifolds of non-vanishing constant curvature are: the de Sitter spacetimes and anti-de Sitter hyperboloids of various radii; the elliptic versions of these; and the various covers of these.[85]

---

[82] The covers of $dS_2(r)$ will play a role in Chapter 8 below.

[83] The isometry groups of these covers are discussed in Monclair, "Isometries of Lorentz Surfaces and Convergence Groups."

[84] As usual: for maximal symmetry, see theorem 3.1 of Patrangenaru, "Lorentz Manifolds"; for isotropy, see theorem 11.6.7 of Wolf, *Spaces of Constant Curvature*.

[85] See Wolf, *Spaces of Constant Curvature*, corollary 11.6.4.

4

Elliptic de Sitter Spacetime

1. Introduction

In Chapter 1 we were guided by the idea that de Sitter spacetime stands to Minkowski spacetime as the sphere stands to the plane. In Chapter 3 we saw that among the surfaces with the local geometry of the two-sphere, there is one, the elliptic plane, that is just as highly symmetric as the sphere—and that the same picture holds in higher dimensions. We also saw that a version of this picture holds in the Lorentzian setting: each de Sitter spacetime has an elliptic rival—and these are the unique maximally symmetric Lorentz manifolds of a given dimension and positive curvature.

This chapter takes a closer look at elliptic de Sitter spacetimes. We are guided by a new analogy:

Elliptic de Sitter: de Sitter :: Elliptic Plane : Sphere.

Section 2 surveys some facts about the Riemannian case, with an emphasis on the reasons for thinking of elliptic spaces as being geometric objects at least as natural as spherical spaces. In Section 3 we turn to our main topic, elliptic de Sitter spacetimes. We again look at some reasons for thinking that, as geometric objects, elliptic de Sitter spacetimes are at least as natural as de Sitter spacetimes. Things are different, however, from the point of view of physics: unlike de Sitter spacetimes, elliptic de Sitter spacetimes are not time-orientable. Indeed, they are probably the most natural example of non-time-orientable Lorentz manifolds without closed timelike curves—and for this reason alone, worth spending some time thinking about.

It is widely held among physicists and philosophers that the time-orientability of de Sitter spacetimes constitutes decisive reason to think them more interesting than their elliptic rivals. This is widely held, but not universally: in Section 4 we will look at some reasons that physicists have given for thinking that elliptic de Sitter spacetimes might provide better models of our world (this is a story that will be completed in the final section of the next chapter). This chapter constitutes something of a digression:

Accelerating Expansion: Philosophy and Physics with a Positive Cosmological Constant. Gordon Belot,
Oxford University Press. © Gordon Belot 2023. DOI: 10.1093/oso/9780192866462.003.0005

ruthless readers immune to the charms of elliptic de Sitter spacetime could skip it without loss of continuity.

2. Elliptic Geometry: The Riemannian Case

What do elliptic spaces look like? Let us begin by being a little more precise about the idea that a given elliptic space is the result of 'identifying' antipodal points in a spherical space of the same dimension and radius.

Fix a spherical space $S = S_n(r)$. Let let $-I : \mathbb{R}_0^{n+1}$ be the antipodal map that sends $x = (x_0, x_1, \ldots, x_n)$ to $-x = (-x_0, -x_1, \ldots, -x_n)$. Note that $(-I)^2$ is the identity map I. For any $x \in S$, we write $\pm x$ for the set $\{x, -x\}$ (the orbit of x under the two-element group generated by $-I$) and we use π to denote the map that sends x to $\pm x$. Standard results about quotient manifolds then imply that if we endow the set

$$\bar{S} := \{\pm x \mid x \in S\}$$

with the topology according to which a subset U of $\bar{S}$ is open if and only if $\pi^{-1}(U)$ is an open subset of S, then $\bar{S}$ is locally homeomorphic to S and inherits from S a Riemannian metric that also renders it locally isometric to S.[1] The resulting Riemannian manifold $\bar{S}$ is of course the elliptic space $\bar{S}_n(r)$. Elliptic n-space is non-orientable for even n, orientable for odd n (see Remark 4.1).

There is a direct relationship between the distance structure of a spherical space and that of the corresponding elliptic space.[2] For any points $x, y \in S$, let us write $d(x, y)$ for the length of the shortest geodesic segment joining x and y in S and $\bar{d}(\pm x, \pm y)$ for the length of the shortest geodesic segment joining $\pm x$ and $\pm y$ in $\bar{S}$. Then:

$$\bar{d}(\pm x, \pm y) = \min\{d(x, y), d(x, -y)\}.$$

Since you can think of points of an elliptic space $\bar{S}$ as pairs of antipodal points in the corresponding spherical space S, you can also think of them as lines through the origin in the ambient $\mathbb{R}_0^{n+1}$ (in which case, the formula above is telling you that the distances between points in the elliptic space

---

[1] See, e.g., O'Neill, *Semi-Riemannian Geometry*, proposition 7.7 and corollary 7.12.
[2] See, e.g., Ratcliffe, *Foundations of Hyperbolic Manifolds*, §2.2.

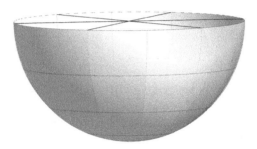

Figure 4.1 The elliptic plane is the result of identifying antipodal points on the boundary of a hemisphere.

are proportional to the smaller of the two angles subtended by the lines representing them).

Here is another way to think of an elliptic space (see Figure 4.1). Let H be a (closed) hemisphere of S (choose a hyperplane through the origin of the ambient Euclidean space and excise all points of S lying to one side of this hyperplane). If x is in the interior of H, then $-x \notin H$; if x is in the boundary of H, then so is $-x$; if x is in the complement of H in S then $-x$ is in the interior of H. So each point in $\bar{S}$ is the image under the projection map $\pi : x \in S \mapsto \pm x \in \bar{S}$ of some point in H. What is more, the restriction of π to the interior of H is a local isometry onto its image. So rather than thinking of $\bar{S}$ as the result of beginning with S and then identifying each point with its antipodal evil twin, we can think of it as the result of beginning with H and then identifying points on the boundary with their antipodal opposite numbers. This is of course easiest to picture in the two-dimensional case, where the boundary of H is just a circle, which projects down to a circle in $\bar{S}$. For $n > 2$, the boundary of H will be an $(n-1)$-dimensional spherical space which will project down to an $(n-1)$-dimensional elliptic space in the n-dimensional elliptic space $\bar{S}$. Note that in each dimension, although H is a manifold-with-boundary, $\bar{S}$ is an ordinary manifold (without boundary)—the boundary of H projects down to a submanifold of the elliptic space $\bar{S}$ that if deleted leaves a space isometric to the interior of H. Of course, care is required when using this representation of elliptic spaces: in the hemisphere model, the pole of the hemisphere and its equator seem special, but we know that in an elliptic space all points are geometrically equivalent to one another, as are all geodesics.

Given their maximal symmetry, there is some temptation to think of the n-spheres and the elliptic n-spaces as equally natural and elegant as mathematical objects, with more or less equal claim to be the basic examples of spaces of constant positive curvature. But many geometers in fact regard the elliptic spaces as being the more natural and interesting objects.[3] To get a sense of their reasons, consider the two-dimensional setting. On a sphere, distinct lines intersect in pairs of antipodal points, and pairs of distinct points can be joined by a single line—unless the points are antipodal, in which case they lie on infinitely many lines. In the elliptic plane the picture is more elegant: each pair of distinct lines intersects in a single point and each pair of distinct points can be joined by a single line.[4] More generally, to many geometers elliptic geometry is more appealing than spherical geometry because it is so closely related to projective geometry.[5] For these and other reasons, in the early years of the twentieth century, the mathematically inclined often took it for granted that given a choice between taking physical space to be spherical and taking it to be elliptic, one would take the elliptic route.[6]

Of course, one could also list some respects in which spherical geometry is 'better' than elliptic geometry.

1) While it is true that for each $n \geq 2$, the isometry groups of n-dimensional spherical space and of elliptic space both have the maximal dimension $n(n+1)/2$, their isometry groups are not the same: $I(\bar{S}_n(r)) \simeq I(S_n(r))/\mathbb{Z}_2$.[7] One *might* think that this constitutes

---

[3] Indeed, in the older literature spherical geometry was often called double elliptic geometry.

[4] In this connection, Hilbert and Cohn-Vossen attribute to spherical geometry a "disturbing property" absent in elliptic geometry (*Geometry and the Imagination*, 237). Thurston complains that "[a]ccounts of spherical geometry are marred by exceptions arising from the existence of antipodes" and therefore absent in elliptic geometry (*Three-Dimensional Geometry and Topology*, Example 1.4.3). And Coxeter praises Klein for having been the first to see "clearly how to rid spherical geometry of its one blemish" (*Non-Euclidean Geometry*, 13).

[5] This is especially salient because of a beautiful account due to Cayley and Klein on which hyperbolic, Euclidean, and elliptic geometry arise via adding structure to projective geometry—see, e.g., Coxeter, *Non-Euclidean Geometry*.

[6] On this and related questions, see Einstein's correspondence in 1916–1918, especially the letter from Schwarzschild of 6 February 1916 (8.188), the letter to Freundlich of 18 February 1917 or later (8.300), the letter to Klein of 26 March 1917 (8.319), the letter from de Sitter of 20 June 1917 (8.355), the letter to de Sitter of 28 June 1917 (8.359), the letter from Klein of 20 March 1918 (8.487), the postcard from Klein of 25 April 1918 (8.518), the letter to Klein of 27 April 1918 (8.523), the letter from Weyl of 19 May 1918 (8.544), the letter to Weyl of 31 June (or May?) 1918 (8.551), and the letter from Klein of 31 May 1918 (8.552). Schwarzschild was interested in this question long before the advent of general relativity—see his "On the Permissible Curvature of Space"; and Schemmel, "The Continuity Between Classical and Relativistic Cosmology in the Work of Karl Schwarzschild." For a modern re-evaluation of these issues, see McInnes, "De Sitter and Schwarzschild-de Sitter According to Schwarzschild and de Sitter."

[7] See fn. 50 of Chapter 3 above.

a sense in which each spherical space $S_n(r)$ is more highly symmetric than the corresponding elliptic space $\bar{S}_n(r)$. In even dimensions one can make a stronger case: the group of isometries of an elliptic space $\bar{S}_n(r)$ is a proper subgroup of the isometry group of the corresponding spherical space $S_n(r)$.[8]

2) Homogeneity and isotropy are the lowest rungs on a ladder of symmetry conditions. A Riemannian n-manifold ($n \geq 2$) is *k-point homogeneous* if an isometry between k-point subsets of the manifold can always be extended to an isometry of the manifold. One-point homogeneity is homogeneity in the usual sense. For Riemannian manifolds, two-point homogeneity is equivalent to isotropy.[9] All and only the hyperbolic, Euclidean, spherical, and elliptic spaces are two-point homogeneous.[10] Spherical, Euclidean, and hyperbolic spaces are k-point homogeneous for each $k = 1, 2, 3, \ldots$.[11] But elliptic spaces are k-point homogeneous only for $k \leq 2$.[12]

REMARK 4.1 (Orientability and Elliptic Geometry). Elliptic n-space is non-orientable for even n, orientable for odd n.[13] To see this, fix an orientation for $\mathbb{E}_{n+1}$ and consider S_n as embedded in $\mathbb{E}_{n+1}$. At each point x of S_n, let N_x be the unit inward-pointing normal to S_n in the tangent space $T_x\mathbb{E}_{n+1}$ (which we identify with $\mathbb{R}^{n+1}$). An ordered basis $(v_1, \ldots, v_n)$ for T_xS_n counts as *positively oriented* if $(N_x, v_1, \ldots v_n)$ matches the fixed orientation for the ambient Euclidean space, otherwise it counts as *negatively oriented*. Let x be a point in S_n, and let $(N_x, v_1, \ldots, v_n)$ be a positively oriented ordered basis for $T_x\mathbb{E}_{n+1}$, so that $(v_1, \ldots, v_n)$ is a positively oriented basis for T_xS_n. Let $-x$ be the image of x under the antipodal map. The antipodal map acts on

---

[8] $O_0(n)$ consists of two connected components, consisting of those orthogonal transformations that preserve orientation and those that reverse orientation. For n even, the antipodal map $x \mapsto -x$ on $\mathbb{R}_0^{n+1}$ is orientation-reversing. It then follows that $I(\bar{S}_n(r))$ is isomorphic to the group of orientation-preserving isometries of $S_n(r)$.

[9] See Wolf, *Spaces of Constant Curvature*, lemma 8.12.1. For a partial analog for Lorentz manifolds, see Beem, "Lorentzian Distance and Curvature," proposition 3.6.

[10] There is some temptation to say: it is only in two-point homogeneous spaces that the notion of a rigid measuring rod capable of free mobility makes sense. But that seems hasty—there is a natural sense in which a torus, being homogeneous and locally isotropic, also supports the existence of rigid, moveable measuring rods (of limited extent). For this history of this question, see Epple, "From Quaternions to Cosmology."

[11] These spaces are in fact *fully homogeneous*: any isometry between subsets can be extended to an isometry of the space. For this result, see Birkhoff, "Metric Foundations of Geometry." For discussion of this and related results, see Belot, *Geometric Possibility*, appendix E.

[12] See Blumenthal, "Congruence and Superposability in Elliptic Space."

[13] For a more general result, see Petersen, *Riemannian Geometry*, §6.3.2.

tangent vectors by reversing their direction. So it sends $(N_x, v_1, \ldots, v_n)$ to $(-N_x, -v_1, \ldots, -v_n)$. Now, $(-N_x, -v_1, \ldots, -v_n)$ is a positively oriented basis for $T_{-x}\mathbb{E}_{n+1}$ if $n + 1$ is even, but not if $n + 1$ is odd. And since the antipodal maps sends N_x to $-N_x$, which is the inward-pointing normal for S_n at $-x$, this means that $(-v_1, \ldots, -v_n)$ is a positively oriented basis for $T_{-x}S_n$ if and only if n is odd, which means that an elliptic n-space is orientable if and only of n is odd.

3. Elliptic Geometry: The Lorentz Case

Fix a de Sitter spacetime $D = dS_n(r)$. Let $-I : \mathbb{R}_1^{n+1} \to \mathbb{R}_1^{n+1}$ be the antipodal map that sends $x = (x_0, x_1, \ldots, x_n)$ to $-x = (-x_0, -x_1, \ldots, -x_n)$. For any $x \in D$ write $\pm x$ for the set $\{x, -x\}$ and use π to denote the map that sends x to $\pm x$. Standard results about quotient manifolds give us a natural way to make the set

$$\{\pm x \mid x \in D\}$$

into a manifold and to equip it with a Lorentz geometry locally isometric to the geometry of D.[14] The result is the elliptic de Sitter spacetime $\bar{D} = \overline{dS}_n(r)$.

We can think of $\bar{D}$ as the collection of antipodal pairs of points on the n-dimensional de Sitter hyperboloid D or as the collection of spacelike lines through the origin in the ambient $(n + 1)$-dimensional Minkowski spacetime. Topologically, an n-dimensional elliptic de Sitter spacetime is a non-trivial $\mathbb{R}$-bundle over an $(n - 1)$-dimensional elliptic space.[15]

Just as in the Riemannian case it is sometimes helpful to think of an elliptic space as a hemisphere with certain identifications made along its boundary, so it is sometimes helpful to think of an elliptic de Sitter spacetime as the result of cutting a de Sitter spacetime in half and then making certain identifications along the boundary.

A *fundamental domain* for the projection map $\pi : x \in D \mapsto \pm x \in \bar{D}$ is a connected open subset U of D (with a suitably regular boundary), such

---

[14] See, e.g., O'Neill, *Semi-Riemannian Geometry*, proposition 7.7 and corollary 7.12.
[15] See, e.g., Calabi and Markus, "Relativistic Space Forms," 72. Note that it follows that elliptic de Sitter spacetimes are not globally hyperbolic, since a globally hyperbolic Lorentz manifold with Cauchy surfaces of topology Σ has topology $\mathbb{R} \times \Sigma$—see, e.g., Wald, *General Relativity*, theorem 8.3.14.

that: (i) U and $-U := -I(U)$ are disjoint; (ii) U and $-U$ have a common boundary, Δ, with $-I(\Delta) = \Delta$; and (iii) $D = U \cup -U \cup \Delta$.

Let U be such a fundamental domain and let V be its closure (so that $V = U \cup \Delta$). The restriction of the projection map π to U is injective but not surjective (each point in $\overline{D}$ is the image of at most one point under π, but those in $\pi(\Delta)$ are the image of none). The restriction of π to V is surjective but not injective (since if $x \in \Delta$, then π maps x and $-x$ to the same point $\pm x$ in $\overline{D}$).

So as with a hemisphere and the corresponding elliptic plane, if U is a fundamental domain for the antipodal map on a de Sitter spacetime D and V is the closure of U, then $\pi(U)$ is an open and dense region of $\overline{D}$ that looks just like U—and $\overline{D}$ as a whole looks like V with certain identifications made along the boundary of U.

It will be helpful to have in mind a couple of ways of constructing fundamental domains for the antipodal map on D. Recall that for any subset S of any spacetime M, the *chronological future*, $I^+(S)$, of S in M is the set of points in M that can be reached by future-directed timelike curves departing from points in S.[16]

i) Let κ be a spacelike hyperplane through the origin of the ambient Minkowski spacetime and let Σ be the intersection of κ with the hyperboloid D. Then Σ is a Cauchy surface of minimal volume and $I^+(\Sigma)$ is a fundamental domain for the antipodal map on D (see Figure 4.2). We will call Σ the *equator* and call $I^+(\Sigma)$ the *upper half* of D.

ii) Alternatively, let κ be a null hyperplane through the origin of the ambient Minkowski spacetime. Then κ splits the de Sitter hyperboloid into two halves, one of which has the form $I^+(\gamma)$ for some timelike geodesic γ, the other of which has the form $I^-(-\gamma)$ for the timelike geodesic $-\gamma$ antipodal to γ, each a fundamental domain for the antipodal map (see Figure 4.3). Following the terminology of Section 2.4 above, we will call $I^+(\gamma)$ the *cosmological patch* of γ.

Choosing a fundamental domain to represent $\overline{D}$ as a region of D (with identifications of antipodal points on the boundary of the region) is a valuable means of visualizing elliptic de Sitter spacetimes. Of course, caution

---

[16] $I^+(S)$ is always an open subset of M—see, e.g., Wald, *General Relativity*, 190.

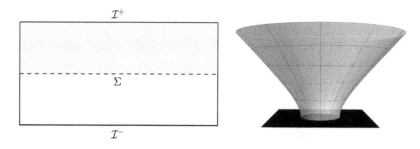

Figure 4.2 The chronological future of a de Sitter Cauchy surface Σ as a fundamental domain. On the left, a Penrose diagram in which the chronological future of the Cauchy surface Σ is shaded. On the right, the region of the de Sitter hyperboloid to the future of Σ. We can picture the elliptic de Sitter spacetime $\overline{D}$ as this future portion, with antipodal points in Σ identified.

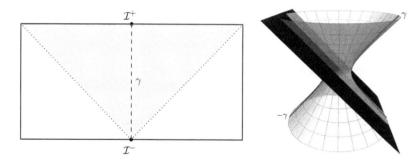

Figure 4.3 The chronological future of a timelike curve as a fundamental domain. On the left, a Penrose diagram in which the chronological future of the timelike geodesic γ is shaded. On the right, the same region as a subset of the de Sitter hyperboloid. We can picture $\overline{D}$ as this region, with antipodal points on the boundary identified.

is required. A choice of fundamental domain will break the underlying symmetry of $\overline{D}$. Elliptic de Sitter spacetimes are isotropic, just as de Sitter spacetimes are; so every timelike geodesic plays that same role in $\overline{D}$ (and likewise for spacelike geodesics). It is also important to note that elliptic de Sitter spacetimes are ordinary manifolds (without boundaries)—the image under the projection map of the boundary of a fundamental domain is *not* a boundary of $\overline{D}$.

De Sitter spacetimes are weird in many ways. Elliptic de Sitter spacetimes are weirder yet. Recall that a Lorentz manifold M is *time-orientable* if there is a continuous way of singling one of the two lobes of the null cone at each point. This is equivalent to the existence of a timelike vector field on M.[17] Elliptic de Sitter spacetimes are not time-orientable (see Exercise 4.1 below).[18]

But elliptic de Sitter spacetime contains no closed causal curves. Suppose that γ is a closed causal curve in $\bar{D}$. Think of $\bar{D}$ as the upper half of the de Sitter hyperboloid, with antipodal points on the equator identified. If γ spends all of its time in the interior of this region, never passing through the equator, then it has the same behaviour as some casual curve in ordinary D— and so cannot be closed. So γ must pass through the equator. So let us assume that $\gamma : [0,2] \to D$ with $\gamma(0) = p$ (a point on the equator), $\gamma(1) = q$ (a point not on the equator), and $\gamma(2) = -p$ (the point on the equator antipodal to p). For $t \in (0,1]$, $\gamma(t)$ must be in the causal future of p (in speaking this way, we exploit the fact that away from the equator, $\bar{D}$ is isometric to a region of our time-oriented D). Similarly, for $t \in [1,2)$, $\gamma(t)$ must be in the causal future of $-p$. So q must be in $J^+(p) \cap J^+(-p)$. But that intersection is empty (see Figure 4.4).

In fact, there is a sense in which timelike geodesics in $\bar{D}$ look just like timelike geodesics in D. Let γ be an inextendible timelike geodesic in D. Then we can think of $\bar{D}$ as the chronological future $I^+(\gamma)$ of γ, together with the boundary of $I^+(\gamma)$, with antipodal points on the boundary identified.

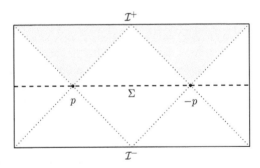

Figure 4.4 The future causal futures of antipodal points in D have no overlap.

---

[17] See, e.g., O'Neill, *Semi-Riemannian Geometry*, lemma 5.32.
[18] Further, even-dimensional elliptic de Sitter spacetimes are space-orientable but not orientable while odd-dimensional elliptic de Sitter spacetimes are orientable but not space-orientable. See Calabi and Markus. "Relativistic Space Forms," theorem 3.

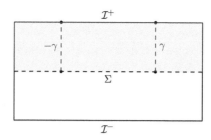

Figure 4.5 A de Sitter Cauchy surface Σ and its chronological future. γ and $-\gamma$ are segments antipodal geodesics. They project down to form a single geodesic in $\bar{D}$.

We can find an open neighbourhood N of γ that is contained entirely in $I^+(\gamma)$—so the result $\bar{N}$ of projecting N down to $\bar{D}$ will be isometric to N.[19] So in particular, the image $\bar{\gamma}$ of γ under the projection map is a geodesic whose affine parameter ranges over all of $\mathbb{R}$. And since $\bar{D}$ is isotropic, we know that the same is true of each timelike geodesic in $\bar{D}$.

How does this fact show up if we are thinking of $\bar{D}$ as the upper half of D (with identifications at the base)? Look at Figure 4.5: the two geodesic segments shown will project down to form a single geodesic in $\bar{D}$. If we think of $\bar{D}$ as the upper half of the de Sitter hyperboloid with antipodal points on the equator identified, then $\pm\gamma$ can be given an affine parameterization according to which as t runs from $-\infty$ to 0, $\pm\gamma$ runs down $+\gamma$ until it reaches a point on the equator; then as t runs from 0 to $+\infty$, $\pm\gamma$ runs up $-\gamma$, starting from the antipodal point on the equator. This representation suggests that there is a natural sense in which $\bar{D}$ has a single boundary at infinity, at which each inextendible causal curve has its endpoints.

Elliptic de Sitter spacetimes stand to de Sitter spacetimes as elliptic spaces stand to spherical spaces. In the Riemannian setting, the relations of incidence between points and lines is neater in the elliptic case than in the spherical case: in elliptic spaces any two distinct points lie on one and only one geodesic, whereas in spherical spaces there is an exception to this rule, with any pair of antipodal points lying on infinitely many geodesics.

Recall from Section 1.2.2 above that in de Sitter spacetime some pairs of distinct points lie on one and only one geodesic, antipodal pairs of points lie

---

[19] So $\bar{D}$ is certainly not compact (since N is not). Indeed, in contrast with the situation for $k \leq 0$, there are no compact Lorentz manifolds of constant curvature $k > 0$—see Klingler, "Complétude des variétés Lorentziennes à courbure constante."

on infinitely many geodesics—and some pairs of points cannot be joined by any geodesic. The picture is prettier in elliptic de Sitter spacetime: each pair of distinct points lies on one and only one geodesic (see Exercise 4.2).[20]

For this and other reasons, elliptic de Sitter spacetime is sometimes regarded as the more elegant and natural geometric object—and for this reason it is preferred to de Sitter spacetime by some geometers.[21]

But from a physical point of view, time-orientable spacetimes are generally preferred to non-time-orientable spacetimes (no matter how elegant). The earliest expression of this view comes in the very letters in which Klein (himself a pioneer of elliptic geometry) brought the elliptic version of de Sitter spacetime to the attention of Einstein and de Sitter.[22] And to this day, it is often taken for granted that temporal orientability is a necessary condition for physical admissibility. The most forceful expression of this view is likely due to Maudlin: "no physicist or philosopher has ever suggested the physical possibility of a temporally non-orientable space-time."[23]

EXERCISE 4.1 (Medium). Show that no elliptic de Sitter spacetime $\overline{D}$ is time-orientable. Explain why the argument you give doesn't also work for elliptic anti-de Sitter spacetime.[24]

EXERCISE 4.2 (Medium). Let $\overline{D}$ be an elliptic de Sitter spacetime and let $\pm p$ and $\pm q$ be distinct points in $\overline{D}$.

a) Show that there is a geodesic in $\overline{D}$ that includes both $\pm p$ and $\pm q$.[25]
b) Show that there is only one such geodesic.[26]

---

[20] So, in particular, elliptic de Sitter spacetimes are geodesically convex. More generally, in dimension three and higher, a complete Lorentz manifold of constant positive curvature is geodesically convex if and only it is not time-orientable—see, e.g, O'Neill, *Semi-Riemannian Geometry*, proposition 9.19.

[21] For earlier advertisements for the superiority of elliptic de Sitter spacetime, see du Val, "Geometrical Note on de Sitter's World," and "On the Discriminations between Past and Future"; and Coxeter, "A Geometrical Background for De Sitter's World."

[22] See fn. 66 above.

[23] Maudlin, *Philosophy of Physics: Space and Time*, 168. For rather different views, see Sklar, *Space, Time, and Space-Time*, 396 f.; and Earman, "What Time Reversal Is and Why It Matters," 257.

[24] HINT: suppose that $\overline{X}$ is a timelike vector field on $\overline{D}$ and consider its lift X to D. For antipodal points $p, -p \in D$, what can you say about $X(p)$ and $X(-p)$? For further background on temporal (non)orientability, see Beem, Ehrlich, and Easley, *Global Lorentzian Geometry*, §3.1. On lifts of vector fields from $\overline{D}$ to D, see O'Neill, *Semi-Riemannian Geometry*, 212.

[25] HINT: let p and q be points in D that project down to $\pm p$ and $\pm q$ in $\overline{D}$ and consider what happens if p and q are connected by no geodesic, exactly one geodesic, or by infinitely many geodesics.

[26] HINT: let $p, -p \in D$ project to $\pm p$ and $q, -q \in D$ project to $\pm q$. Consider the Minkowski tangent plane π determined by p, q, and the origin.

EXERCISE 4.3 (Medium). Above it was noted that if we select the upper half of dS as our fundamental domain, then it is natural to regard $\overline{dS}$ as having a single ideal boundary, on which all inextendible timelike curves have their endpoints. What if we instead work with the cosmological patch of an observer γ in dS as our fundamental domain? Do we end up with the same picture or a different one?

EXERCISE 4.4 (Harder). Let $\overline{D}$ be an elliptic de Sitter spacetime and let $\gamma_1, \ldots,$ γ_N be timelike curves in $\overline{D}$. Suppose that each γ_i is parameterized by proper time and defined for all $t \in \mathbb{R}$. Suppose further that while distinct γ_i and γ_j may intersect, they never overlap for non-zero intervals of proper time. Let us say that $\Gamma = \{\gamma_1, \ldots, \gamma_N\}$ is *consistent* if for any $p \in \overline{D}$ at which distinct γ_i and γ_j intersect, the tangent vectors to γ_i and γ_j lie in the same lobe of the null cone in the tangent space at p.

a) For $N = 3$, construct a consistent Γ and an inconsistent Γ.
b) Claim:

> Given an inconsistent Γ, it is sometimes possible to construct a consistent Γ', with $|\Gamma| = |\Gamma'|$ and where every γ in Γ' is either in Γ or the time-reverse of a curve in Γ.

Show that this claim is true. Does every inconsistent Γ correspond to a consistent Γ' in this way?

4. And Yet...

In a course of lectures on cosmology, Schrödinger explores what he calls *the elliptic interpretation* of de Sitter spacetime, which "consists in deeming antipodes to represent the same world-point or event."[27] Schrödinger hastens to show that closed causal curves are impossible under the elliptic interpretation.

> This removes the *prima facie objections* to the intended identification of the antipodes. But there is a circumstance which, to my view, positively suggests it. It is this. The potential experience of an observer who moves on an arbitrary timelike world line from the infinite past ... to the infinite

---

[27] Schrödinger, *Expanding Universes*, 7.

future...includes just one half of $[dS]$. Which half depends on the world line he follows.[28]

Schrödinger goes on to make a case that elliptic de Sitter spacetime is more physically reasonable than standard de Sitter spacetime. The entire discussion is fascinating, but here I will quote just one further portion, infused with a bracing Hegelian verificationism.

> it does seem rather odd that two or more observers, even such as 'sat on the same school bench' in the remote past, should in future, when they have 'followed different paths in life,' experience different worlds, so that eventually certain parts of the experienced world of one of them should remain *by principle* inaccessible to the other and vice versa. It is true that they can never become aware of this being so, except by theoretical considerations such as we are conducting just now, certainly not by comparing notes. For whenever they met for this purpose, they have the same forecone and thus have at that moment reached the same potential experience. Still, we—or at any rate some of us—are used to regard the 'real world around us' as a mental construct based on the community experience of all normal, sane persons. From this point of view, one may find it distasteful to accept a world model according to which two observers who separate are likely to have the possible sharing of their experience stopped with regard to some parts of it that are interesting and relevant to them respectively. Indeed, this will happen sooner or later, unless they reach the distant future at precisely the same azimuth. One way of avoiding this is to accept the elliptic interpretation.[29]

Schrödinger's interest in elliptic de Sitter spacetime may have been idiosyncratic—and the reasons he gave for his interest even more so. But over the years there has been recurring interest among physicists in Schrödinger's elliptic interpretation (sometimes for reasons not wholly unrelated to those given by Schrödinger).[30]

---

[28] Schrödinger, *Expanding Universes*, 9.   [29] Schrödinger, *Expanding Universes*, 12.

[30] There are some dozens of papers in this vein—not a large number by the standards of theoretical physics, but not a negligibly small one either. Some influential contributions: Gibbons, "The Elliptic Interpretation of Black Holes and Quantum Mechanics"; Sánchez, "Quantum Field Theory and the 'Elliptic Interpretation' of de Sitter Spacetime"; Friedman and

A couple of features of elliptic de Sitter spacetime have driven recent interest. One caught Schrödinger's eye: for any worldline, every point in the spacetime can be connected to some point on that worldline by a causal curve. The other is the fact (mentioned above) that elliptic de Sitter spacetimes can be thought of as having a single spacelike boundary on which each inextendible causal curve has its endpoints (in contrast to the de Sitter case, where causal curves begin at past infinity and end at future infinity). In order to put ourselves in a position to appreciate why these features are intriguing to contemporary physicists, we will need to undertake a long detour through some fascinating terrain. We will return to the allures of elliptic de Sitter spacetime in the final section of the next chapter.

QUESTION 4.1 (Medium?). Schrödinger held that the question of truth does not typically arise for scientific models, since in order for a representation to be capable of truth, it must be the sort of thing that can be directly compared to reality.[31] Is this view sustainable? Does it help to make sense of his claims about the elliptic interpretation?

QUESTION 4.2 (Medium?). Schrödinger notes that temporally irreversible laws, such as the second law of thermodynamics, are at least *prima facie* incompatible with temporally non-orientable spacetimes.

> Still this may not be fatal to the elliptic interpretation. For it is well known that the irreversible laws of thermodynamics can only be based on the statistics of microscopically reversible systems on condition that the statistical theory be autonomous in defining the arrow of time. If any other laws of nature determine this arrow, the statistical theory collapses.... From this point of view, a reversible world model is highly desirable.[32]

Higuchi, "Quantum Field Theory in Lorentzian Universes from Nothing"; Parikh, Savonije, Verlinde, "Elliptic de Sitter Space"; Banks and Mannelli, "De Sitter Vacua, Renormalization, and Locality"; Aguirre and Gratton, "Inflation without a Beginning"; and Hackl and Neiman, "Horizon Complementarity in Elliptic de Sitter Space."

[31] See Schrödinger, *Science and Humanism*, 21 ff. For further discussion, see D'Agostino, "The *Bild* Conception of Physical Theory."

[32] Schrödinger, *Expanding Universes*, 14.

What does Schrödinger have in mind here? The ellipsis in the above quotation corresponds to a reference to an earlier paper of Schrödinger's.[33] How does Schrödinger's argument in that paper work—and what is its bearing on his contention here about the elliptic interpretation?

QUESTION 4.3 (Smaller?). Can we make sense of time-asymmetric laws in elliptic de Sitter spacetime?

---

[33] Schrödinger, "Irreversibility."

5

The Anti-Hero

1. Introduction

In this chapter we take a ramble through the $\Lambda < 0$ realm, examining the geometry of anti-de Sitter spacetime and its closest relatives. We begin in Section 2 with a review of some basic facts about the geometry of the anti-de Sitter hyperboloid and its universal cover, anti-de Sitter spacetime. Section 3 is devoted to a study of the conformal completion of anti-de Sitter spacetime. This attention to anti-de Sitter geometry may seem quixotic: given that observation seems to speak in favour of a positive value for the cosmological constant, it might be thought that anti-de Sitter geometry should be of little relevance to contemporary physics. That, it turns out, is very far from the truth. Indeed, the anti-de Sitter case is generally held to provide profound insight into the process of black hole evaporation—and much else—via the *Anti-de Sitter/Conformal Field Theory Correspondence* (AdS/CFT). That story is the subject of Section 5, with Section 4 reviewing some background to it. Section 6 ties up some loose ends and introduces some others. The $\Lambda < 0$ detour we take in this chapter is not a mere diversion: anti-de Sitter geometry, the technology of conformal completion, and the AdS/CFT correspondence will each play prominent roles in Chapters 6–9 below.

2. Anti-de Sitter Basics

For convenience, we will specialize to the case of four spacetime dimensions (essentially all of the discussion below carries over to other dimensions with minimal modifications).[1] Recall that the anti-de Sitter hyperboloid $\underline{AdS}_4(r)$ is the subspace of $\mathbb{R}^5_2$ determined by the condition

---

[1] For helpful overviews, see: Hawking and Ellis, *The Large Scale Structure of Space-Time*, §5.2; Griffiths and Podolský, *Exact Space-Times in Einstein's General Relativity*, chapter 5; Moschella, "The de Sitter and Anti-de Sitter Sightseeing Tour"; and Bonsante and Seppi, "Anti-de Sitter Geometry and Teichmüller Theory," part 1.

Accelerating Expansion: Philosophy and Physics with a Positive Cosmological Constant. Gordon Belot, Oxford University Press. © Gordon Belot 2023. DOI: 10.1093/oso/9780192866462.003.0006

$$-x_0^2 - x_1^2 + x_2^2 + x_3^2 + x_4^2 = -r^2.$$

If we introduce the coordinates $(\tau, \rho, \theta, \varphi)$ on $\underline{AdS}_4(r)$ via

$$x_0 = r \cosh \rho \sin(\tau/r)$$
$$x_1 = r \sinh \rho \cos \theta$$
$$x_2 = r \sinh \rho \sin \theta \cos \varphi$$
$$x_3 = r \sinh \rho \sin \theta \sin \varphi$$
$$x_4 = r \cosh \rho \cos(\tau/r)$$

with $\tau \in [0, 2\pi r)$, $\rho \in [0, \infty)$, $\theta \in [0, \pi]$, and $\varphi \in [0, 2\pi)$, then the anti-de Sitter metric is

$$ds^2 = -\cosh^2 \rho \, d\tau^2 + r^2 \left[d\rho^2 + \sinh^2 \rho \left(d\theta^2 + \sin^2 \theta \, d\varphi^2 \right) \right].$$

This metric is time-independent and has spatial sections with hyperbolic geometry (the expression within square brackets being the line element for hyperbolic three-space).[2]

Anti-de Sitter hyperboloids are strange beasts. For one thing, as noted in Section 3.5, they feature closed timelike curves (indeed, all timelike geodesics in $\underline{AdS}_n(r)$ have this feature). For present purposes, those are largely a distraction. So alongside $\underline{AdS}_4(r)$, let us also consider its universal cover, the anti-de Sitter spacetime $AdS_4(r)$, the result of 'unrolling' the time coordinate of $\underline{AdS}_4(r)$: the metric for $AdS_4(r)$ is given by the same expression as the metric for $\underline{AdS}_4(r)$ above, but now with τ ranging over the entire real line.[3] $AdS_4(r)$ has the same local geometry as $\underline{AdS}_4(r)$ but has the topology of $\mathbb{R}^4$. Whereas in $\underline{AdS}_4(r)$, time consists of a single cycle of length $2\pi r$, in $AdS_4(r)$ it consists of a two-way infinite succession of chunks of length $2\pi r$. Anti-de Sitter spaceime is maximally symmetric (and so also isotropic and homogeneous).[4]

---

[2] Due to the ρ-dependence of the $d\tau^2$ term, it is a *warped product* (rather than a product *simpliciter*) of a one-dimensional Lorentz manifold with hyperbolic space. On warped products, see, e.g., O'Neill, *Semi-Riemannian Geometry*, chapter 7.

[3] Similarly, the metric of the k-fold covering $AdS_4^{(k)}(r)$ is given by the same expression as that of $\underline{AdS}_4(r)$, but with $\tau \in [0, 2\pi k)$.

[4] $\underline{AdS}_n(r)$ is a quotient of the simply connected $AdS_n(r)$ by a freely and properly acting countable group of isometries. So the result noted in fn. 50 of Chapter 3 above implies that $\dim I(AdS_n(r)) \geq \dim I(\underline{AdS}_n(r))$. So $AdS_n(r)$ must be maximally symmetric, since $\underline{AdS}_n(r)$ is. Note, further, that theorem 11.6.7 of Wolf, *Spaces of Constant Curvature* tells us

$AdS_4(r)$ contains no closed timelike curves—but, like $\underline{AdS}_4(r)$, it fails to be globally hyperbolic. The problem is a strange feature shared by both spacetimes. They are geodesically complete. But for any spacelike slice Σ it is possible to find a point p not on Σ and an inextendible null geodesic through p that doesn't meet Σ—light signals can 'escape to infinity' without registering on Σ; so Σ cannot be a Cauchy surface.[5]

This behaviour is a reflection of the fact that in anti-de Sitter spacetimes, conformal infinity is timelike. Let's approach this point in small steps.

i) Recall, first, that the metric for the round three-sphere, $S_3(1)$, can be written in the form:

$$d\Omega_3^2 = d\bar{\chi}^2 + \sin^2 \bar{\chi} \left(d\bar{\theta}^2 + \sin^2 \bar{\theta} d\bar{\varphi}^2 \right),$$

where $\bar{\chi} \in [0, \pi]$, $\bar{\theta} \in [0, \pi]$, and $\bar{\varphi} \in [0, 2\pi)$. $\bar{\chi} = 0$ and $\bar{\chi} = \pi$ correspond to opposite poles of the three-sphere. Except for these poles, each $\bar{\chi} =$const. section of the three-sphere is a two-sphere, with the volume growing as $\bar{\chi}$ moves away from 0 and π, and reaching a maximum at the 'equator' given by $\bar{\chi} = \pi/2$. So if we impose the restriction $\bar{\chi} \in [0, \pi/2)$, then we end up with a hemi-three-sphere: a manifold with the topology of $\mathbb{R}^3$ carrying a metric with constant curvature $k = 1$, and whose boundary in the ambient sphere has topology S^2. Let us denote this space $\frac{1}{2}S_3$.

ii) Recall, next, that the four-dimensional Einstein static universe, E_4, has topology $\mathbb{R} \times S^3$ and metric:

$$ds^2 = -d\bar{t}^2 + \left[d\bar{\chi}^2 + \sin^2 \bar{\chi} \left(d\bar{\theta}^2 + \sin^2 \bar{\theta} d\bar{\varphi}^2 \right) \right],$$

where $\bar{t} \in \mathbb{R}$, $\bar{\chi} \in [0, \pi]$, $\bar{\theta} \in [0, \pi]$, and $\bar{\varphi} \in [0, 2\pi)$. The term in square brackets is just the metric for $S_3(1)$. If we impose the coordinate restriction $\bar{\chi} \in [0, \pi/2)$, then we end up with a spacetime in which the $\bar{t} =$const. spatial sections each have the geometry and topology

that the elliptic anti-de Sitter spacetime $\overline{AdS}_n(r)$ and the various k-fold coverings $\underline{AdS}_n^{(k)}(r)$ of anti-de Sitter spacetime are quotients of $AdS_n(r)$ by normal subgroups of its group of isometries. So the result quoted in fn. 50 of Chapter 3 implies that $\overline{AdS}_n(r)$ and the $\underline{AdS}_n^{(k)}(r)$ are maximally symmetric (because it tells us that their isometry groups have the same dimension as that of anti-de Sitter spacetime).

[5] This feature of anti-de Sitter spacetimes is exploited in Earman and Norton, "Forever Is a Day."

of $\frac{1}{2}S_3$. Let us denote this spacetime $\frac{1}{2}E_4$. Note that the boundary of $\frac{1}{2}E_4$ in E_4 is very simple: the intersection of this boundary with a $\bar{t}$ =const. slice is just an $S_2(1)$. So the boundary of $\frac{1}{2}E_4$ in the ambient E_4 is a copy of E_3, the three-dimensional Einstein static universe.

iii) Turning now to $AdS_3(r)$, note that if we introduce coordinates χ and t related to ρ and τ

$$\tan \chi = \sinh \rho$$
$$t = \tau/r$$

then the metric of $AdS_4(r)$ becomes:

$$ds^2 = \frac{r^2}{\cos^2 \chi} \left[-dt^2 + d\chi^2 + \sin^2 \chi \left(d\theta^2 + \sin^2 \theta d\varphi^2 \right) \right],$$

where $t \in \mathbb{R}$, $\chi \in [0, \pi/2)$, $\theta \in [0, \pi]$, and $\varphi \in [0, 2\pi)$. Given that $\chi \in [0, \pi/2)$, the expression in square brackets is the metric for $\frac{1}{2}E_4$. So the entire expression tells us that $AdS_4(r)$ is conformally isometric to $\frac{1}{2}E_4$, from which it follows that the boundary $\mathcal{I}$ of $AdS_4(r)$ can be identified with E_3 (see Figure 5.1). Note that due to the timelike nature of $\mathcal{I}$, for any point in $AdS_4(r)$, every null geodesic 'escapes to infinity' and fails to register on sufficiently late surfaces of constant time (see Figure 5.2).[6]

Timelike geodesics never pull this sort of disappearing act in anti-de Sitter spacetimes. Intuitively, because a negative cosmological constant acts as an attractive force, escape velocity for massive bodies is infinite in anti-de Sitter spacetimes. But timelike anti-de Sitter geodesics do a weird thing of their own. Let us introduce yet another set of coordinates for $AdS_4(r)$ via:

$$x_0 = r \sin(\hat{t}/r)$$
$$x_1 = r \cos(\hat{t}/r) \sinh \hat{\rho} \cos \hat{\theta}$$
$$x_2 = r \cos(\hat{t}/r) \sinh \hat{\rho} \sin \hat{\theta} \cos \hat{\varphi}$$
$$x_3 = r \cos(\hat{t}/r) \sinh \hat{\rho} \sin \hat{\theta} \sin \hat{\varphi}$$
$$x_4 = r \cos(\hat{t}/r) \cosh \hat{\rho}.$$

---

[6] Could we avoid this result by treating the boundary as physical and imposing reflecting boundary conditions? Yes. We will eventually see that doing so would have interesting consequences.

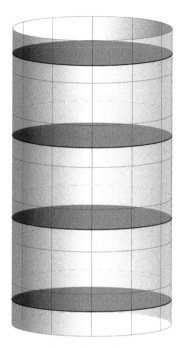

Figure 5.1 A conformal completion of anti-de Sitter spacetime. The boundary has the structure of the d-dimensional Einstein static universe. The interior is conformally isometric to the $(d+1)$-dimensional anti-de Sitter spacetime (selected time-slices of the interior are highlighted here). The diagram should be infinitely extended both upwards and downwards.

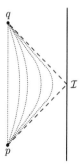

Figure 5.2 Causal geodesics in anti-de Sitter spacetime. Two null geodesics are shown as dashed lines: one leaves p and reaches null infinity, the other reaches q from null infinity. Also depicted via dotted lines is the behaviour of timelike geodesics departing from p.

Then the metric for $AdS_4(r)$ becomes:

$$ds^2 = -d\hat{t}^2 + r^2 \cos^2(\hat{t}/r) \left[d\hat{\rho}^2 + \sinh^2 \hat{\rho} \left(d\hat{\theta}^2 + \sin^2 \hat{\theta} d\hat{\varphi}^2 \right) \right],$$

where $\hat{t} \in \mathbb{R}$, $\hat{\rho} \in [0, \infty)$, $\hat{\theta} \in [0, \pi]$, and $\hat{\varphi} \in [0, 2\pi)$ (note that these coordinates do *not* cover all of $AdS_4(r)$). Curves along which $\hat{\rho}$, $\hat{\theta}$, and $\hat{\varphi}$ are constant are timelike geodesics orthogonal to the $\hat{t}$ =const. spatial slices. Note that timelike geodesics departing from a given point on the $\hat{t} = -r\pi/2$ slice will initially diverge—then reconverge to the point with the same spatial coordinates at $t = r\pi/2$, from which they will then diverge, only to reconverge at the point with the same spatial coordinates at $t = 3r\pi/2$. And so on. See Figure 5.2 for behaviours characteristic of null and of timelike geodesics in anti-de Sitter spacetimes.

EXERCISE 5.1 (Easier). *Silberstein spacetime* is $\mathbb{R}^4$ equipped with the line element

$$ds^2 = -d\tau^2 + r^2 \left[d\rho^2 + \sinh^2 \rho \left(d\theta^2 + \sin^2 \theta d\varphi^2 \right) \right],$$

where $\tau \in \mathbb{R}$, $\rho \in [0, \infty)$, $\theta \in [0, \pi]$, and $\varphi \in [0, 2\pi)$. (This is the anti-de Sitter metric without the initial ρ-dependent redshift factor in front of $d\tau^2$.) Silberstein took this to be a Lorentz space of constant negative curvature.[7] Show that he was incorrect.

QUESTION 5.1 (Larger). Suppose that we had an empirically adequate model of our world set in an anti-de Sitter hyperboloid $\underline{A}$. Then we could also make empirically adequate models set in the double cover $\underline{A}^{(2)}$ of $\underline{A}$, or in the triple cover $\underline{A}^{(3)}$ of $\underline{A}$, ..., or in the universal cover A of $\underline{A}$ by concatenating two, three, ..., or countably infinitely many copies of the history that occurs in our basic model. On a straightforward realist reading, these models will differ about how many perfect duplicates of you occur in world history. On a Reichenbachian reading, each of these models is just a different way of describing the same possibility—and there is no fact about which of these

---

[7] Silberstein, "General Relativity without the Equivalence Hypothesis," §2. For more on Silberstein, arguably the preeminent early frenemy of relativity, see Havas, "The General-Relativistic Two-Body Problem and the Einstein-Silberstein Controversy"; and Flin and Duerbeck, "Silberstein, Relativity, and Cosmology."

models best represents that possibility.[8] On a Leibnizean reading, there is again just one possibility here, best represented by $\underline{A}$.[9] A sociological claim: the realist reading tends to be more popular among contemporary philosophers, the Reichenbachian and Leibnizean readings more popular among contemporary physicists.

a) Make the strongest case you can in favour of a Reichenbachian or Leibnizean approach.[10]

b) Explore connections between this question and the interpretation of dualities in physics.[11]

3. The Boundary of *AdS*

Let M be a manifold and let g and h be Lorentz metrics on M. Then we say that g and h are *conformally equivalent* if there is a positive function Ω on M such that $g = \Omega^2 h$. Every metric is conformally equivalent to many others: if g is a Lorentz metric on M and Ω is a positive function on M, then g and $\Omega^2 g$ are conformally equivalent. We write $[g]$ for the class of metrics on M conformally equivalent to g.

Above we have seen that de Sitter spacetime and anti-de Sitter spacetime are each conformally equivalent to portions of the Einstein static universe. We again specialize to four spacetime dimensions (what follows can easily be transposed to other dimensions). Then the four-dimensional Einstein static universe E_4 has topology $\mathbb{R} \times S^3$ and metric given by:

$$ds^2 = -dt^2 + d\chi^2 + \sin^2\chi \left(d\theta^2 + \sin^2\theta d\phi^2\right),$$

where $t \in (-\infty, \infty)$, $\chi \in [0, \pi]$, $\theta \in [0, \pi]$, and $\phi \in [0, 2\pi)$.

[8] As I understand him, van Fraassen is committed to this approach: see §III.4 of his *Introduction to the Philosophy of Time and Space*; and Belot, "Transcendental Idealism among the Jersey Metaphysicians."

[9] See, e.g., Grünbaum, *Philosophical Problems of Space and Time*, §7.A. For further discussion and references, see van Fraassen, *Introduction*, §III.1.b; and Hacking, "The Identity of Indiscernibles."

[10] Here it might or might not help to resort to the neo-verificationism of Peacocke, "The Limits of Intelligibility."

[11] For a philosophical introduction to this literature, see Huggett and Wüthrich, "Out of Nowhere: Duality." For a survey of some interpretative options, see Le Bihan and Read, "Duality and Ontology."

We have seen that we can choose coordinates so that the metric of the de Sitter space $dS_4(r)$ becomes:

$$ds^2 = \frac{r^2}{\sin^2 t} \left[-dt^2 + d\chi^2 + \sin^2 \chi \left(d\theta^2 + \sin^2 \theta d\phi^2 \right) \right],$$

with the coordinate ranges as for E_4, except that $t \in (0, \pi)$. So we can identify de Sitter spacetime with a temporally finite chunk of the Einstein static universe—and the de Sitter and Einstein static metrics on this chunk are conformally equivalent. Since conformally equivalent metrics agree about causal structure, we can investigate the causal structure of $dS_4(r)$ by investigating the causal structure of a part of the Einstein static universe.

And we have seen that we can choose coordinates so that the metric of $AdS_4(r)$ becomes:

$$ds^2 = \frac{r^2}{\cos^2 \chi} \left[-dt^2 + d\chi^2 + \sin^2 \chi \left(d\theta^2 + \sin^2 \theta d\phi^2 \right) \right],$$

with the coordinate ranges as for E_4, except that $\chi \in [0, \pi/2)$. So we can identify anti-de Sitter spacetime with 'one half' of the Einstein static universe (a temporally infinite piece in which a closed three-hemisphere has been excised from each spatial slice). The anti-de Sitter and Einstein static metrics on this region are conformally equivalent. So we can investigate the causal structure of AdS by examining the causal structure of our one-half Einstein static universe.

The magic of these constructions is that they allow us to replace the study of infinitely extended geodesics in dS and AdS with the study of finite geodesic segments in the Einstein static universe: the conformal factor maps the infinite to the finite.[12]

Now, when it comes to our temporally finite chunk of the Einstein static universe, all timelike and null geodesics that are inextendible within this region can be assigned endpoints on the boundary $\mathcal{I}$ of this region, corresponding to the hypersurfaces $t = 0$ and $t = \pi$ in the full Einstein static universe. So it is natural to think of $\mathcal{I}$ as an ideal boundary that we could add to de Sitter spacetime—consisting of the 'points at infinity' that are the ideal endpoints of inextendible timelike and null de Sitter geodesics.

---

[12] This way of putting things is a bit misleading: a curve that is a non-null geodesic relative to one metric may not be a geodesic relative to conformally equivalent metrics. But it will do for present purposes.

Similarly, in the AdS case, all inextendible spacelike and null geodesics on the half Einstein static universe can be assigned endpoints on the boundary $\mathcal{I}$ of this region—and this boundary can likewise be thought of as consisting of the 'points at infinity' that are the ideal endpoints of inextendible spacelike and null AdS geodesics.

Let us pursue this way of thinking a little further. In the AdS case, what can we say about the structure of $\mathcal{I}$? It is a three-dimensional manifold with topology $\mathbb{R} \times S^2$ that inherits from its embedding in the Einstein static universe a Lorentz metric:

$$ds^2 = -dt^2 + \left(d\theta^2 + \sin^2\theta d\phi^2\right),$$

where $t \in (-\infty, \infty)$, $\theta \in [0, \pi]$, and $\phi \in [0, 2\pi)$. Since the spatial part of the metric is just the metric of the round two-sphere, this is the metric of the three-dimensional Einstein static universe.

So there is a sense in which the ideal boundary of AdS on which spacelike and null geodesics end has the structure of the (three-dimensional) Einstein static universe. But, it turns out, this is not the most interesting way of thinking of this boundary—it pays to be a little more abstract. Let (M, g) be a Lorentz manifold. A *conformal completion* for (M, g) is a Lorentz manifold-with-boundary (N, h) with non-empty boundary $\mathcal{I}$, a diffeomorphism $d \colon M \to N/\mathcal{I}$ (that we use to identify M and $N/\mathcal{I}$), and a non-negative function Ω on N such that:

    i) on M, $h(x) = \Omega^2 g(x)$ (so that Ω is positive on $N/\mathcal{I}$);
    ii) on $\mathcal{I}$ we have $\Omega(x) = 0$ and $d\Omega(x) \neq 0$.[13]

In this case we call $\mathcal{I}$ a *conformal boundary* of (M, g). The first requirement in the definition tells us that on $N/\mathcal{I}$, the metrics g and h are conformally equivalent. The second tells us that the boundary $\mathcal{I}$ represents conformal infinity for (M, g)—inextendible null geodesics in $N/\mathcal{I}$ have endpoints on $\mathcal{I}$ but cannot reach them for finite values of their affine parameters relative to g.[14]

This definition does a good job of capturing the notion of a spacetime's admitting a well-behaved conformal boundary. But note that if (M, g),

[13] This is an adaptation (allowing $\Lambda \neq 0$) of a definition of Geroch, "Asymptotic Structure of Space-Time."

[14] On this point, see, e.g., Valiente Kroon, *Conformal Methods in General Relativity*, §7.1.

(N, h), d, and Ω are as in the definition of a conformal completion and ω is any positive-valued function on N, then (M, g), $(N, \omega^2 h)$, d, and $\omega \cdot \Omega$ and will also be as in the definition. In other words, whenever we have one representative $(\mathcal{I}, h_\mathcal{I})$ of the geometry of the conformal boundary of (M, g), we can make another $(\mathcal{I}, (\omega^2 \cdot h)|_\mathcal{I})$ by multiplying h by any positive function. So it is really the conformal equivalence class $[h|_\mathcal{I}]$ that encodes the intrinsic geometry of $\mathcal{I}$ *qua* conformal boundary of M, rather than any particular member of it.

Our discussion above showed that AdS_4 admits a conformal boundary $\mathcal{I}$ with topology $\mathbb{R} \times S^2$. One way to represent the geometry of that boundary is with the Lorentz metric of the Einstein static universe. But this specific metric structure, we can now see, depends on the details of the construction of $\mathcal{I}$. A more natural, intrinsic, and revealing description is that $\mathcal{I}$ has the geometry of the conformal equivalence class of the Einstein static universe.

A conformal equivalence class of metrics may seem like a funny sort of object. The three-dimensional Einstein static universe has a four-dimensional group of isometries (time translations and spatial rotations). But some metrics conformally equivalent to the Einstein static metric have trivial isometry groups (just pick an irregular conformal factor). What should we say about the symmetry group of the conformal equivalence class of the Einstein static metric?

We will need to introduce a notion of symmetry more general than isometry in this context if we want each representative of the conformal equivalence class to have the same symmetries. A *conformal isometry* of a Lorentz manifold (M, g) is a diffeomorphism $d : M \to M$ such that g and $d^* g$ are conformally equivalent. Every metric is conformally equivalent to many others (positive functions are plentiful), but only special metrics admit conformal isometries—just as every Lorentz manifold is isometric to many others (diffeomorphisms are plentiful), but only special metrics admit isometries.[15] But if Lorentz metrics g_1 and g_2 on M are conformally

---

[15] In general, one expects that for any differential equation, solutions with minimal symmetry are in some sense generic—for this theme, see Olver, *Applications of Lie Groups to Differential Equations*. So one expects that generic solutions of the Einstein equations should not admit any isometries. This expectation is fulfilled: see Marsden and Isenberg, "A Slice Theorem for the Space of Solutions of Einstein's Equations"; and Andersson, "Momenta and Reduction for General Relativity." In four spacetime dimensions, except for Lorentz spaces of constant curvature and certain plane wave solutions, any conformal isometry of a vacuum solution (with or without Λ) is in fact an isometry. See Eardley *et al.*, "Homothetic and Conformal Symmetries of Solutions to Einstein's Equations"; and Garfinkle and Tian, "Spacetimes with Cosmological

equivalent, then a given diffeomorphism $d : M \to M$ is a conformal isometry of both or of neither—conformally equivalent metrics admit exactly the same conformal isometries.[16]

So it is natural to think of the symmetries of a conformal equivalence class of metrics as being the conformal isometries of its members. What about the case that interests us? The three-dimensional Einstein static universe admits a four-dimensional group of isometries and a ten-dimensional group of conformal isometries. So the isometry group of AdS_4 is the same size as (=has the same dimension as) the conformal isometry group of the three-dimensional Einstein static universe that encodes the conformal geometry of the boundary the conformal completion of AdS_4. More is true: the two groups are isomorphic.[17]

More generally, every infinitesimal isometry of a Lorentz manifold that admits a conformal completion induces an infinitesimal symmetry of the corresponding conformal boundary—but it is *not* in general the case that there are only as many conformal isometries of the boundary as there are isometries of the bulk.[18] Indeed, the geometry of the conformal boundary of Minkowski spacetime (which is represented by degenerate metrics rather than by Lorentz metrics) has an infinite-dimensional group of symmetries (the Bondi-Metzner-Sachs group)—more on this in Section 6.3 below.

4. Observer Complementarity

Recall some of the puzzles surrounding Hawking radiation and black hole evaporation.[19] In the mid-1970s, Hawking shocked the world by showing it to be a consequence of quantum field theory on a Schwarzschild background

Constant and a Conformal Killing Field Have Constant Curvature." (Note that if (M,g) is a vacuum solution, we do not in general expect metrics conformally equivalent to g to be vacuum solutions.)

[16] See, e.g., Stephani *et al.*, *Exact Solutions of Einstein's Field Equations*, §35.4.3.

[17] See, e.g., Frances, "The Conformal Boundary of Anti-de Sitter Spacetimes," §2. The group in question is an infinite-sheeted cover (but *not* the universal cover) of the isometry group of AdS_4. It is not a matrix Lie group. For an explicit characterization, see Eshkobilov, Musso, and Nicolodi, "On the Restricted Conformal Group of the $(1+n)$-Einstein Static Universe."

[18] See, e.g., Valiente Kroon, *Conformal Methods*, 10.4.2.

[19] The following discussion is of necessity both highly selective (mentioning only a few relevant topics and a vanishingly small fraction of the relevant literature) and highly superficial (just sketching—or gesturing at—the relevant considerations). Readers interested in more depth should consult the excellent recent reviews: Harlow, "Jerusalem Lectures on Black Holes and Quantum Information"; and Wallace, "Why Black Hole Information Loss Is Paradoxical."

that an observer far from a black hole should detect thermal radiation with temperature inversely proportional to the mass of the black hole—and by going on to argue that if an isolated massive body collapses to form a black hole, then that black hole should gradually evaporate (lose mass), until it finally disappears, leaving behind only the emitted thermal radiation.[20] Hawking then shocked the world again by observing that even if the matter whose collapse led to the formation of the black hole was initially in a pure state, the Universe would be in a mixed state once the black hole had evaporated away. So the standard quantum framework is not adequate to describe such a process from beginning to end: in that framework, time evolution is implemented by unitary operators; but no unitary operator can map a pure state to a mixed state.[21]

The following twenty-five or so years saw a great deal of controversy, with some physicists content to accept Hawking's conclusion, but with most eager to find a way to reconcile unitary evolution with the phenomenon of Hawking radiation.[22] One of the reasons for all the excitement was the hope that understanding the fate of black holes would allow progress to be made on the project of a quantum theory of gravity. The obverse of this hope was that absent a quantum theory of gravity, it was impossible to model in detail the fate of a black hole—if for no other reason, then because an evaporating black hole should eventually reach Planck mass, at which point quantum field theory on curved spacetime should certainly no longer be a reliable guide. So there was considerable room for speculation about exotic ways that unitarity might be preserved. Perhaps the end result of evaporation includes not just thermal radiation, but also an exquisite tiny object with enough degrees of freedom so that the total post-evaporation state of the Universe (radiation plus remnant) is in fact pure. Or perhaps evaporation is total, but results in some new sort of singularity, which cuts off the history of the Universe (so that there is no post-evaporation state in which all is thermal radiation). And so on. But the most tantalizing possibility was that Planck-scale corrections to Hawking's calculations would imply that the radiation emitted by a black hole was only approximately thermal—leaving room to maintain that evaporation is total but also unitary, resulting in a final state

---

[20] Hawking, "Particle Creation by Black Holes." For a textbook treatment from a mathematical physics perspective, see Wald, *Quantum Field Theory on Curved Spacetimes and Black Hole Thermodynamics*.

[21] Hawking, "Breakdown of Predictability in Gravitational Collapse."

[22] For an overview from towards the end of this period, see Belot, Earman, and Ruetsche, "The Hawking Information Loss Paradox."

that encodes in the emitted radiation every detail of the initial pre-collapse state of the matter from which the black hole was formed.

However, this suggestion faces an immediate difficulty. It is a basic fact about quantum mechanics—essentially just a consequence of the linearity of the dynamics—that there can be no mechanism that duplicates quantum states (see Exercise 5.2). In particular, the horizon of a black hole cannot be a quantum xerox machine. Suppose that a system collapses to form a black hole that then evaporates completely, leaving nothing but radiation behind. After the system has fallen through the horizon of the black hole there may be sufficient information behind the horizon to allow complete reconstruction of the system's initial state or there may be sufficient information external to the horizon to allow such reconstruction—but this information cannot be available in both regions at the same time. But in general relativity, the crossing of the horizon of a large black hole is typically an unremarkable event—so it is hard to believe that someone who crosses the horizon with the collapsing system wouldn't have just as much information available post-crossing as pre-crossing. So some fancy footwork awaits anyone tempted to maintain that full information about the original system should be available in the post-evaporation radiation. For it seems that if this information were gradually encoded in the radiation as the black hole evaporated, then even at early stages some portion of the information must have been duplicated: in the quantum state defined on a spatial slice that includes portions of the region exterior to the black hole horizon as well as regions behind that horizon we ought to be able to find two copies of the information needed for a reconstruction of (some aspects of) the initial state.

One of the most striking and influential suggested responses to this problem is the principle of *black hole complementarity* enunciated by Susskind, Thorlacius, and Uglum.[23] These authors observe that it follows from the linearity of quantum mechanics that if the pre-collapse state and the post-evaporation state are both pure, then the information necessary to reconstruct the initial state must remain outside the horizon and be unavailable to an infalling observer:

> In other words, all distinctions between initial states of infalling matter must be obliterated before the state crosses the global event horizon.

---

[23] For what follows, see Susskind, Thorlacius, and Uglum, "The Stretched Horizon and Black Hole Complementarity," 3744 and 3760.

But, as they note, this is in tension with common sense:

> Although we shall not introduce specific postulates about observers who fall through the global event horizon, there is a widespread belief which we fully share. The belief is based on the equivalence principle and the fact that the global event horizon of a very massive black hole does not have large curvature, energy density, pressure, or any other invariant signal of its presence. For this reason, it seems certain that a freely falling observer experiences nothing out of the ordinary when crossing the horizon.

The resolution of this difficulty, they suggest, requires an understanding of the proper ambitions of physics.

> The assumption of a state...which simultaneously describes both the interior and the exterior of a black hole seems suspiciously unphysical. Such a state can describe correlations which have no operational meaning, since an observer who passes behind the event horizon can never communicate the result of any experiment performed inside the black hole to an observer outside the black hole. The above description...can only be made use of by a "superobserver" outside our Universe. As long as we do not postulate such observers, we see no logical contradiction in assuming that a distant observer sees all infalling information returned in Hawking-like radiation, and that the infalling observer experiences nothing unusual before or during horizon crossing. Only when we try to give a combined description, with a standard quantum theory valid for both observers, do we encounter trouble. Of course, it may be argued that a quantum field theoretic description of gravity dictates just such a description, whether we like it or not. If this is the case, such a quantum field theory is inconsistent with our postulates; therefore, one or the other is incorrect.

An observer who falls through the horizon initially sees nothing strange—and may have access to full information about the initial state of the collapsing system. From the perspective of an observer who stays far outside the black hole horizon, all information about the collapsing system is stored in a stretched horizon (a membrane just outside the horizon) where it is available to be encoded in Hawking radiation. Crucially, no observer is ever in a position to combine information about the collapsing system collected from behind the horizon with information about the collapsing system encoded in Hawking radiation.[24]

> In many respects, the situation seems comparable to that of the early part of the century. The contradictions between the wave and the particle theories of light seemed irreconcilable, but careful thought could not reveal any logical contradiction. Experiments of one kind or the other revealed either particle or wave behavior, but not both. We suspect that the present situation is similar. An experiment of one kind will detect a quantum membrane, while an experiment of another kind will not. However, no possibility exists for any observer to know the results of both. Information involving the results of these two kinds of experiments should be viewed as *complementary* in the sense of Bohr.

For present purposes, there are two key ideas here. We should be satisfied if each perspective supports a coherent story without aspiring to coherently reconciling these stories. And in this special case, at least, full information about the system of interest is available from both perspectives considered.

Black hole complementarity was both influential and controversial when it was first introduced—and it remains so.[25] But it invited generalization. A paper by Parikh, Savonije, and Verlinde made the natural connection with Schrödinger's elliptic interpretation and catalyzed interest in elliptic de Sitter spacetime.

> The arguments that led to black hole complementarity can also be applied to other types of event horizons, in particular to cosmological event horizons [such as the boundary of an observer's causal future in dS]. A better name would therefore be "observer complementarity." In its strongest

---

[24] This is immediate for an observer who remains outside the horizon. More delicate arguments are required to show that any observer who collects information about the collapsed object from Hawking radiation then falls through the horizon will be unable to receive signals from observers who passed though the horizon with the collapsing body. See Susskind and Thorlacius, "Gedanken Experiments Involving Black Holes"; and Hayden and Preskill, "Black Holes as Mirrors."

[25] For sympathetic philosophical assessments, see van Dongen and de Haro, "On Black Hole Complementarity"; and Muthukrishnan, "Unpacking Black Hole Complementarity." For the most influential and controversial criticism of black hole complementarity, see Almheiri *et al.*, "Black Holes: Complementarity or Firewalls?"

form, it postulates that each observer has complete information, and can in principle describe everything that happens within his/her cosmological horizon using pure states. This information may appear to different observers in different—complementary—guises: one observer may pass smoothly through the horizon, whereas another observer may see there a source of hot radiation. Although these drastically different realities may seem to be inconsistent, it is important to recognize that paradoxes arise only when one takes the unphysical perspective of a global superobserver.[26]

As we will see below, interest in observer complementarity and horizons in de Sitter-like spacetimes continues to be a focus of interest—but (for the time being, at least) interest has shifted from cosmological patches to static patches.

EXERCISE 5.2 (Easier). Let $\mathcal{H}$ be a Hilbert space and ϕ_0 be a unit-norm vector in $\mathcal{H}$. A *quantum xerox machine* is a unitary operator $U : \mathcal{H} \otimes \mathcal{H} \to \mathcal{H} \otimes \mathcal{H}$ such that any unit-norm vector $\psi \in \mathcal{H}$, $U(\psi \otimes \phi_0) = \psi \otimes \psi$ (think of ϕ_0 as a blank piece of paper, ψ as the state you would like to copy, and U as the dynamics induced by a quantum xerox machine). Show that there can be no such U.[27]

EXERCISE 5.3 (Easier). Let $\mathcal{H}_1$ and $\mathcal{H}_2$ be the Hilbert spaces associated with two physical systems, so that quantum states of the composite system that they form are represented by unit-norm vectors in $\mathcal{H} := \mathcal{H}_1 \otimes \mathcal{H}_2$. So properties of the composite system correspond to closed linear subspaces of $\mathcal{H}$. One property the composite system could have (or lack) is being in an entangled state (with respect to the decomposition into our given systems). What linear subspace of $\mathcal{H}$ corresponds to this property?[28]

5. AdS/CFT

Back to our main thread. The debate about black hole evaporation information loss underwent a sea change in the late 1990s.[29]

---

[26] Parikh, Savonije, and Verlinde "Elliptic de Sitter Space," 1.
[27] Or see Wooters and Zurek, "A Single Quantum Cannot be Cloned" or Dieks, "Communication by EPR Devices."
[28] For a philosophical application, see Weinstein, "Undermind."
[29] Warning: although the story I am about to rehearse is the standard one I have heard it dismissed by a prominent participant—"You are probably just getting that from the literature."

A harbinger was 't Hooft's argument (building on earlier insights of Bekenstein and of Hawking) that the physical degrees of freedom inside a region containing a black hole correspond to a finite-dimensional Hilbert space whose size is a function, not of the volume of the region, but of the area of its boundary—from which 't Hooft concluded that a quantum theory of gravity should obey the principle (now usually called the *holographic principle*) that "given any closed surface, we can represent all that happens inside it by degrees of freedom on this surface itself."[30]

The decisive development was Maldacena's string-theoretic implementation of the holographic principle in his conjecture of a duality between the asymptotically anti-de Sitter sector of quantum gravity and a conformally invariant quantum field theory living on the boundary of anti-de Sitter spacetime.[31] The conjectured *AdS/CFT correspondence* (which need not be understood in string-theoretic terms) is now accepted as established across large swathes of the physics community on the basis of an impressive number of successful tests (despite the fact the quantum gravity half of the correspondence remains a moving target).[32]

For present purposes, the key point is that the boundary theory is a perfectly ordinary quantum field theory in which time evolution is implemented by a unitary operator: so if a pure state is given along a time-slice of the boundary, then dynamical evolution gives us a pure state on any other time-slice of the boundary. In some quarters, this is taken to settle the question of information loss.

> To create the black hole, we can act with the CFT creation operators ... to create an infalling spherical shell of matter that from the bulk point of view is expected to collapse into a black hole. We can then evolve this state forward in the CFT and see what it looks like after a time which is greater than the bulk evaporation time. This evolution is unitary, so to the extent that AdS/CFT is a definition of the bulk theory, this resolves the information problem in the sense of telling us the answer: information is preserved.[33]

---

[30] 't Hooft, "Dimensional Reduction in Quantum Gravity," 289.

[31] See Maldacena, "The Large N Limit of Superconformal Field Theories and Supergravity." The characterization of the correspondence given below derives from subsequent explications and amplifications of Maldacena's original approach.

[32] For a review, see Hubeny, "The AdS/CFT Correspondence." Of course, one might also take the view that in the absence of a quantum theory of gravity (and of a specification of what *asymptotically anti-de Sitter* means in the quantum context), the correspondence has the status of a conjecture that has yet to be so much as clearly stated—for this attitude, see, e.g., Unruh and Wald, "Information Loss," §5.3.

[33] Harlow, "Jerusalem Lectures," §VI.H.

Or, more briefly:

> By discovering the AdS/CFT correspondence...Maldacena definitively
> answered the question of whether information can escape from a black
> hole. It can.[34]

We have ventured into deep waters which we cannot possibly plumb
adequately here. But before pressing on, let me offer a couple of points of
clarification that may be helpful to philosophical readers tempted to venture
further.

i) We don't know what the asymptotically anti-de Sitter sector of quantum gravity looks like because we don't yet have a theory of quantum
gravity. But in practice, we can often get by with a characterization
of the asymptotically anti-de Sitter sector of general relativity. To
this end, note, first, that under mild assumptions on the behaviour
of matter, the conformal boundary of a $\Lambda < 0$ general relativistic
spacetime is timelike.[35] Let (M, g) be a four-dimensional spacetime
that arises as a solution of some reasonable $\Lambda < 0$ Einstein-matter
equations.

> We say that (M, g) is *weakly asymptotically anti-de Sitter* if it admits
> a conformal completion and its stress-energy tensor satisfies suitable
> fall-off conditions at conformal infinity.

> We say (M, g) is *strongly asymptotically anti-de Sitter* if it is weakly
> asymptotically anti-de Sitter and its conformal boundary has the
> topology and conformal geometry of the Einstein static universe.[36]

---

[34] Penington, "Entanglement Wedge Reconstruction and the Information Paradox," §1.

[35] See Penrose, "Zero Rest Mass Fields Including Gravitation," §11.

[36] For this basic approach (and for the relevant fall-off conditions), see Ashtekar and Magnon,
"Asymptotically Anti-de Sitter Space-Times." For a variant that covers the higher-dimensional
case, see Ashtekar and Das, "Asymptotically Anti-de Sitter Spacetimes." See also the equivalent
approach of Henneaux and Teitelboim, "Asymptotically Anti-de Sitter Spaces," which makes
explicit the sense in which the geometry of a strongly asymptotically anti-de Sitter spacetime
approaches the geometry of anti-de Sitter spacetime at spatial infinity.
Note that whereas when $\Lambda \geq 0$, well-behaved static and complete solutions of the vacuum
Einstein equations are flat (which implies that in fact $\Lambda = 0$), when $\Lambda < 0$, there are well-
behaved static and complete vacuum solutions that are asymptotically anti-de Sitter but not
locally anti-de Sitter. See, on the one hand, Anderson, "On Stationary Vacuum Solutions to
the Einstein Equations"; and Chen, "On Stationary Solutions to the Vacuum Einstein Field
Equations"; and, on the other hand, Anderson, Chruściel, and Delay, "Non-trivial, Static,
Geodesically Complete, Vacuum Space-Times with a Negative Cosmological Constant. I and II."

Because the conformal boundary of a weakly asymptotically anti-de Sitter spacetime is timelike, one might have expected there to be flux of 'conserved' quantities such as energy across $\mathcal{I}$ in generic non-stationary weakly asymptotically anti-de Sitter spacetimes (the corresponding phenomenon occurs in the $\Lambda = 0$ regime).[37] But, in fact, in a vacuum weakly asymptotically de Sitter spacetime whose conformal boundary admits infinitesimal conformal isometries, so that 'conserved quantities' are well-defined, these quantities are absolutely conserved and there is no flux across $\mathcal{I}$ (although matter fields propagating on such spacetime *can* exhibit net flux of energy through $\mathcal{I}$ under these boundary conditions).[38] And this of course remains true in a strongly asymptotically anti-de Sitter spacetime (where the group of conformal isometries of the boundary is isomorphic to the ten-dimensional group of isometries of AdS_4).

ii) A related point. Recall that in anti-de Sitter spacetime null geodesics exhibit a sort of behaviour completely different from anything possible in Minkowski spacetime or de Sitter spacetime: a future-directed null geodesic that departs from a point p will not intersect spacelike slices sufficiently far to the future of p—it will instead 'escape to infinity'. So it might seem that a radiating isolated body in an (asymptotically) anti-de Sitter spacetime is analogous to an ordinary open system—in particular, its energy should not be constant, so it should be modelled using a time-dependent Hamiltonian. But we want the dual CFT to be an ordinary quantum field theory with a time-independent Hamiltonian. So most treatments of AdS/CFT impose reflective boundary conditions for the bulk theory—these restrict the possible conformal geometries of the boundary (requiring conformal flatness) and rule out energy flux across $\mathcal{I}$ (whether or not matter is present).[39] This

---

[37] For a discussion of the time-dependence of 'conserved' quantities like the Bondi mass in the $\Lambda = 0$ setting, see, e.g., Wald, *General Relativity*, §11.2.

[38] See Ashtekar and Magnon, "Asymptotically Anti-de Sitter Space-Times," L41 f. Note that the conformal boundary of a merely weakly asymptotically anti-de Sitter spacetime can have any conformal geometry consistent with its topology—in particular, such a conformal boundary may admit no conformal isometries.

[39] For important exceptions, see Almheiri *et al.*, "The Entropy Bulk of Quantum Fields and the Entanglement Wedge of an Evaporating Black Hole"; and Penington, "Entanglement Wedge Reconstruction." Note that in in four spacetime dimensions, a weakly asymptotically anti-de Sitter spacetime with conformal boundary of topology $\mathbb{R} \times S^2$ is strongly asymptotically anti-de Sitter if and only if reflective boundary conditions hold—see Ashtekar and Magnon, "Asymptotically Anti-de Sitter Space-Times"; and Hawking, "The Boundary Conditions for Gauged Supergravity." For an investigation of consistent boundary conditions for fields on

makes it possible to have so-called *large* AdS black holes that are in equilibrium with the heat bath of their own radiation.

6. AdS/CFT and Elliptic de Sitter Spacetime

Let us turn at last to what all this has to do with elliptic de Sitter spacetime. One would naturally like to find a $\Lambda > 0$ analog of AdS/CFT, since it is not obvious which aspects of the analysis of the $\Lambda < 0$ sector can be expected to carry over to the physically relevant $\Lambda > 0$ case. This has proved to be an intractable problem. One immediate obstacle is that whereas a conformal boundary for anti-de Sitter spacetime is itself represented by a Lorentz manifold, so that the boundary provides a setting for a quantum field theory with unitary dynamics, the conformal boundary of de Sitter spacetime looks all wrong—it is spacelike and it comes in two pieces. Here again are Parikh, Savonije, and Verlinde:

> Let us briefly review the puzzles that arise in conventional de Sitter space. We have already mentioned observer complementarity. Another issue is that of holography. We would like to have a holographic dual description of gravity for all of the various asymptotic geometries. Recently, we have learned to describe string theory in spacetimes that asymptotically approach an anti-de Sitter geometry. The AdS conformal field theory (CFT) correspondence is by now well established, and in principle gives a nice holographic description of string theory in these backgrounds. . . . But de Sitter space requires . . . another type of holography, because there is no spatial or null infinity. Various authors have argued that it should be a kind of timelike holography, for which the holographic screens are spacelike surfaces in the asymptotic past or future of global de Sitter space. . . . A somewhat confusing aspect of holography in global de Sitter space, however, is that it has two disconnected boundaries. If we think of the dual CFT as living on these boundaries, then we have to somehow compute correlation functions of operators some of which may be inserted on one

---

a fixed anti-de Sitter background, see Ishibashi and Wald, "Dynamics in Non-Globally-Hyperbolic Static Spacetimes. III." For a treatment of the initial-boundary problem for the vacuum Einstein equations in the asymptotically anti-de Sitter setting subject to boundary conditions more general than reflective boundary conditions, see Friedrich, "Einstein Equations and Conformal Structure." For more on the significance of the choice of AdS boundary conditions, see Section 6.8 below.

boundary, while others may act on another boundary. Not only is it unclear how to compute such correlation functions, it is also unclear what their physical interpretation is.[40]

Part of their motive for working with elliptic de Sitter spacetime was the thought that a model of $\Lambda > 0$ physics with a single boundary might provide a better starting point for the project of finding the de Sitter analog of AdS/CFT.

This idea did not pan out, as it happens. Indeed, twenty years on, there is still no satisfactory $\Lambda > 0$ analog of AdS/CFT. In the meantime, something else interesting has happened. In the first enthusiasm over AdS/CFT, it was a natural thought that the correspondence called for spacetimes with single boundaries. But, of course, even in the $\Lambda = 0$ regime, the most basic model of an eternal black hole is the maximal extension of the Schwarzschild solution, which has two asymptotic regions connected by a wormhole, each with its own conformal boundary.[41] And when $\Lambda < 0$, the most basic model of an eternal black hole is the corresponding Schwarzschild-anti-de Sitter solution, which likewise has two asymptotic regions and two conformal infinities (each with the same structure as the conformal boundary of anti-de Sitter spacetime).[42] In another influential paper, Maldacena explored two models for the physics of an eternal $\Lambda < 0$ black hole from the AdS/CFT perspective.[43] (i) A single-boundary model in which the bulk spacetime is an elliptic version of the Schwarzschild-anti-de Sitter solution, which features a single conformal boundary.[44] (ii) A two-boundary model in which the bulk spacetime is the standard Schwarzschild-anti-de Sitter spacetime with its double conformal infinity, so that the theory on the boundary lives on two separate copies of the conformal boundary of anti-de Sitter spacetime. It is the second model that has dominated subsequent discussion.[45] This is largely because it has a feature that brings to the fore one of the most fascinating aspects of AdS/CFT: an approximately Schwarzschild-anti-de Sitter bulk

[40] Parikh, Savonije, and Verlinde "Elliptic de Sitter Space," 1.

[41] See, e.g., O'Neill, *Semi-Riemannian Geometry*, chapter 13; or Griffiths and Podolský, *Exact Space-Times*, §8.3.

[42] For details, see, e.g., Griffiths and Podolský, *Exact Space-Times*, §9.4.3.

[43] Maldacena, "Eternal Black Holes in Anti-de Sitter."

[44] For earlier investigations of elliptic versions of black hole spacetimes, see, e.g., Rindler, "Elliptic Kruskal-Schwarzschild Space"; Gibbons, "The Elliptic Interpretation of Black Holes and Quantum Mechanics"; and Louko and Marolf, "Inextendible Schwarzschild Black Hole with a Single Exterior."

[45] But see 't Hooft, "Black Hole Unitarity and Antipodal Entanglement."

geometry corresponds to a CFT state in which the states of the two conformal boundaries are entangled with one another.[46] But now we are again in danger of straying into deep waters indeed, as will be made clear by the following commentary on Maldacena's analysis of the boundary state corresponding to the Schwarzschild-anti-de Sitter black hole.

> The proposal of Maldacena is very natural but has dramatic implications. The individual terms in the superposition... are product states in a system of two non-interacting CFTs. In these states, the two theories have absolutely nothing to do with one another, so in the gravity picture, these states must correspond to two completely separate asymptotically AdS spacetimes. On the other hand, the quantum superposition of these states... apparently corresponds to the extended black hole, where the two sides of the geometry are connected by smooth classical spacetime in the form of a wormhole. The remarkable conclusion... is that by taking a specific quantum superposition of disconnected spacetimes, we obtain a connected spacetime.... Alternatively, we can say that by entangling the degrees of freedom underlying the two separate gravitational theories in a particular way, we have glued together the corresponding geometries![47]

QUESTION 5.2 (Larger?). Exercise 5.3 tells us that there is no quantum observable that takes one value (with certainty) if a joint system is in an entangled state and takes another value (with certainty) if the joint system is in an unentangled state—entanglement is not a measurable property. On the picture just mentioned, is it supposed to follow that spacetime topology is likewise not a measurable property?[48]

[46] As Maldacena notes, this insight was prefigured in earlier investigations—see "Eternal Black Holes," 4.

[47] Van Raamsdonk, "Lectures on Gravity and Entanglement," 309 f.—this review is a good place to begin, for anyone interested in the mysteries lumped together under the heading *ER=EPR*.

[48] On this problem, see §5.1 of Bain, "The RT Formula and Its Discontents."

6

Asymptotically de Sitter Spacetimes

1. Prologue

Of course, in our Universe there appears to be a positive cosmological constant—so anti-de Sitter geometry is of no direct physical relevance. This is not to say that our world has de Sitter geometry or even locally de Sitter geometry. But, if we understand things aright, we can be confident that its geometry will become more and more similar to de Sitter geometry in the distant future: this appears to be a more or less inevitable consequence of a positive cosmological constant, under a wide range of conditions. So we henceforth turn our attention towards geometries that are in some sense asymptotically de Sitter. In this chapter, we will look at one natural way of making sense of the notion of an asymptotically de Sitter geometry, in terms of conformal completion. In the next chapter we will review some of the senses in which de Sitter geometry is an attractor within the $\Lambda > 0$ sector of general relativity. In the final two chapters, we will draw out some consequences of life in a world with a de Sitter-like future.

Let's begin, by way of a warmup exercise, with an example that illustrates an important fact that stands behind much of the discussion of this chapter (and of this book as a whole): adding a term to a differential equation can radically change the qualitative behaviour of solutions—even if the additional term depends on a constant that we can take to be arbitrarily small.

We can illustrate this fact with a pair of equations that provide toy models of a sort of behaviour common in physically realistic models of fluid dynamics.[1]

The *inviscid Burgers equation* on $\mathbb{R}^2$ is:

$$\partial_t u(t,x) + u(t,x) \cdot \partial_x u(t,x) = 0.$$

---

[1] For an introductory treatment, see the discussion of Olver, *Introduction to Partial Differential Equations*, §§2.3 and 8.4. For full accounts, see Smoller, *Shock Waves and Reaction-Diffusion Equations*; or Dafermos, *Hyperbolic Conservation Laws in Continuum Physics*. Note the equations we will discuss are associated with Ehrenfest's student Burgers, and so are *Burgers equations* or *Burgers' equations*—but definitely not *Burger's equations*.

Accelerating Expansion: Philosophy and Physics with a Positive Cosmological Constant. Gordon Belot, Oxford University Press. © Gordon Belot 2023. DOI: 10.1093/oso/9780192866462.003.0007

This is a first-order non-linear hyperbolic partial differential equation. *First-order* because it involves only one derivative in t. *Non-linear* because it is not in general true that the sum or difference of two solutions is a solution (blame the multiplicative structure of equation's second term). *Hyperbolic*, in the sense that it is suited to describe wave phenomena.[2] Here it is to be understood that waves have a finite speed of propagation, so that in systems governed by such equations, the initial state in one region has an immediate effect only on the states of those regions that are its immediate neighbours, and thus only eventually comes to play a role in determining the states of distant regions.

We can think of the waves described by this equation as propagating along the x-axis with speed $u(x,t)$—so the amplitude of a wave determines its speed of propagation. It follows that a wave with large positive amplitude can overtake a wave initially to its right but with a smaller amplitude. When this happens shock waves form and the equation breaks down (because u is no longer differentiable).[3] Only special initial data (with u never decreasing as x increases) lead to shock-free solutions—typical smooth initial data can be evolved only for a finite time before singularities emerge.[4]

As the reader may shrewdly suspect, there is also a *viscid Burgers equation*:

$$\partial_t u(t,x) + u(t,x) \cdot \partial_x u(t,x) = \kappa \partial_{xx} u(t,x),$$

(here $\kappa > 0$ is the *viscosity*). This is also a first-order non-linear partial differential equation. But it is parabolic rather than hyperbolic, which is to say that it shares many qualitative features with the heat equation and other equations describing diffusion processes. Such equations (in contrast to the hyperbolic case) model processes in which disturbances have infinite speed of propagation—the heat equation, for instance, implies that if the temperature of an infinite iron bar is initially positive in some finite

---

[2] Any textbook on partial differential equations will feature a precise characterization of hyperbolic, parabolic, and elliptic equations—although the characterization offered will vary depending on the level of generality that the author is aiming for. For the sort of informal characterization used here, see Hadamard, "Les problèmes aux limites dans la théorie des équations aux dérivées partielles," 207.

[3] For a brief introduction to shock waves, see Keyfitz, "Shocks." For details, see the references of fn. 1 above.

[4] For bounded initial data, weak solutions (continuous but not differentiable) defined for all $t > 0$ exist. But uniqueness fails—so that further input is required to select the physically relevant solution corresponding to a given initial data set.

region and zero elsewhere, then it is non-zero everywhere at all subsequent times.

In fact, the *Hopf-Cole transformation* sets up a duality between solutions of the viscid Burgers equation and positive solutions of the heat equation: if $w(x,t) > 0$ solves the heat equation $\partial_t w = \kappa \partial_{xx} w$, then

$$u(x,t) := -2\kappa \frac{\partial_x w(x,t)}{w(x,t)}$$

solves the viscid Burgers equation (and every solution of the viscid Burgers equation arises in this way).

In the case of the heat equation, initial data satisfying mild fall-off conditions at spatial infinity determine smooth solutions defined for all $t > 0$ (indeed, evolution via the heat equation instantaneously *increases* the degree of smoothness of such solutions, so there is no danger of the equation becoming inapplicable because u ceases to be differentiable). The Hopf-Cole transformation guarantees that the corresponding solutions of the viscid Burgers equation have the same feature. So adding a viscosity term $\kappa \partial_{xx} u$ to the right hand side of the inviscid Burgers equation make a profound qualitative difference to the behaviour of solutions—no matter how small κ is.[5]

In this chapter, we will look at one sense in which the qualitative behaviour of solutions of the Einstein field equations depends on whether a cosmological constant term is included in those equations—we will see that there are deep divides between the cases of a positive cosmological constant (no matter how close to zero), a vanishing cosmological constant, and a negative cosmological constant (no matter how close to zero). Intuitively speaking, a geometry should count as asymptotically de Sitter/Minkowski/anti-de Sitter if, as you travel further and further in any direction along certain geodesics, the metric approaches that of de Sitter/Minkowski/anti-de Sitter spacetime (at a suitable rate).[6] The technology of conformal completion provides an elegant way of implementing this intuitive idea in the case where we are interested in null geodesics.

---

[5] It is nonetheless possible to use the so-called *method of vanishing viscosity* to characterize the physically acceptable weak solutions of the inviscid Burgers equations mentioned in fn. 4 above—this involves taking the $\kappa \to 0$ limit of families of solutions to the viscid Burgers equation.

[6] This is a sufficient condition but not a necessary one—we will also want to include certain solutions in which some geodesics of interest vanish into singularities.

In Section 5.3 above, we reviewed the characterization of asymptotically anti-de Sitter spacetimes in terms of the structure of their conformal completions. Here we will review parallel accounts of asymptotically de Sitter and asymptotically Minkowski spacetimes. Intuitively speaking, we seek to understand the global structures of (asymptotically) empty spacetimes by using conformal transformations to replace 'infinitely long' null geodesics by 'finitely long' null geodesics, so that we can attach an ideal boundary to spacetime representing the 'endpoints' of the worldlines of photons.[7] We have already seen that the conformal boundary of anti-de Sitter spacetime is timelike (i.e., it can be represented by a Lorentz manifold) and that the conformal boundary of de Sitter spacetime is spacelike (i.e, it can be represented by a pair of Riemannian manifolds). Here we will see that the conformal boundary of Minkowski spacetime is null (i.e., its geometry can be represented by a degenerate metric). If a vacuum solution admits a conformal completion, the geometric nature (timelike, spacelike, or null) of the boundary is determined by the sign of Λ ($\Lambda < 0$, $\Lambda > 0$, or $\Lambda = 0$).[8] So it is natural to explicate the intuitive notion of an asymptotically de Sitter/Minkowski/anti-de Sitter geometry in terms of the structure of the conformal completion of the geometry, when possible.

We begin in the next section by considering some methodological remarks of Penrose and of Geroch concerning what one can hope to learn by deploying the technology of conformal completion. This is followed by examinations of the notions of asymptotically Minkowski and asymptotically de Sitter spacetimes within the framework of conformal completion. The picture of geometry and physics that emerges differs strikingly depending on the sign of the cosmological constant.

QUESTION 6.1 (Larger!). As discussed in Chapter 5 above, the AdS/CFT correspondence is the focus of much investigation of the physics of black holes. How confident should we be that lessons learned about black holes in the $\Lambda < 0$ context will carry over to the physically relevant $\Lambda > 0$ case? Could this be one of the areas where we find qualitatively different behaviour, depending on the sign of Λ?

---

[7] More carefully: replacing null geodesics whose affine parameters range over unbounded sets of real numbers by null geodesics whose affine parameters range over bounded sets of real numbers.

[8] See Penrose, "Conformal Treatment of Infinity," lecture II.

2. The Rules of the Game

Before plunging in, it will be worthwhile to think a little bit about the point of studying the conformal completion of various Lorentz geometries that arise in general relativity.

Consider, first, some remarks of Penrose, focused on the $\Lambda = 0$ regime, where *asymptotically flat* is a commonly used variant on *asymptotically Minkowski*.

> Asymptotically flat spacetimes are interesting, not because they are thought to be realistic models for the entire universe, but because they describe the gravitational fields of isolated systems, and because it is only with asymptotic flatness that general relativity begins to relate in a clear way to many of the important aspects of the rest of physics, such as energy, momentum, radiation, etc.[9]

As Penrose goes on to say: one restricts attention to spacetimes that are asymptotically flat at spatial infinity (i.e., in which initial data along Cauchy surfaces asymptotically approach trivial Minkowski initial data as one proceeds outwards along spacelike geodesics) in order to make sense of mass and angular momentum; and one restricts attention to spacetimes that are asymptotically flat at null infinity (those that approach Minkowski geometry as one proceeds outwards along null geodesics) in order to make sense of the distinction between systems which emit gravitational radiation and those which do not, in order to model the detection of gravitational waves, and so on.[10] Penrose introduced the framework of conformal completions of spacetimes in order to provide a geometric encapsulation of the complex picture of the asymptotic behaviour of gravitational radiation that emerged in the 1950s and early 1960s in the work of, amongst others, Bondi, Sachs, and Penrose himself.[11]

---

[9] Penrose, "Some Unsolved Problems in Classical General Relativity," 632.

[10] For an especially illuminating development of both approaches, and a discussion of the relation between them, see Ashtekar, Bombelli, and Reula, "The Covariant Phase Space of Asymptotically Flat Gravitational Fields."

[11] For non-technical overviews of these developments, see Kennefick, *Traveling at the Speed of Thought,* chapters 7–9; and Frauendiener, "Conformal Infinity—Development and Applications," §2.

Consider, also, some incisive remarks of Geroch bearing on the desirability of a notion of an isolated system in general relativity and concerning criteria for fixing such a notion.

> It is in a sense only through a suitable notion of an isolated system that one acquires any ability at all to deal individually with various subsystems in the Universe—in particular, to assign to subsystems such physical attributes as mass, angular momentum, character of emitted radiation, etc. . . . In any case, the standard procedure within a given theory is first to obtain a more or less precisely defined class of solutions "representing isolated systems" within the mathematical framework of the theory, and then to attribute to solutions within that class various properties of possible physical interest.[12]

Geroch goes on to note that in general relativity this project turns out to be more difficult than usual, for the same reason that most things are more difficult than usual in general relativity: while in other theories there is a sharp distinction between the dynamical fields and the geometry against which they evolve, in general relativity the metric is both the stage and one of the players. Still, even in general relativity, there is a natural starting point: the intuition that in a situation in which only a single massive body exists, spacetime geometry should be asymptotically approximately Minkowskian far from that body.

> The first step, then, is to translate "isolated system" to mean "asymptotically flat." Recognition of this step is, however, far from solving the problem, for what is "asymptotically flat" to mean? Presumably, something such as "far from the matter, the metric of spacetime approaches some flat metric," or "far from the matter, the Riemann tensor of space-time approaches zero." Unfortunately, the notion of one tensor field's approaching another on a manifold does not in general make sense. Comparison of components with respect to some chart, for example, will not do, for the result of the comparison will in general depend crucially on the choice of chart.[13]

Penrose's idea of attaching a conformal boundary to a spacetime is a means to resolves this difficulty.

[12] Geroch, "Asymptotic Structure of Space-Time," 1 f.
[13] Geroch, "Asymptotic Structure," 2.

Appropriate asymptotic behavior of the metric g_{ab} is expressed in terms of the behavior of Ω and g_{ab} near the points at infinity; that of other physical fields on the space-time in terms of their behavior near these same points. The precise conditions to be imposed on all these fields turns out to be a rather delicate issue. Conditions too strong will have the effect of eliminating solutions which would seem clearly to represent isolated systems; conditions too weak may have the effect of admitting too many solutions or, what is worse, may result in a structure which is so weak that potentially useful aspects of the asymptotic behavior of one's fields are lost in a sea of bad behavior.[14]

In Newtonian physics, life is easy—one has little choice except to say that the gravitational potential of a compact isolated body falls off as $1/r$. Things are not so neat in general relativity.

There are no "correct" or "incorrect" definitions, only more or less useful ones. It is perfectly possible that there turn out to be a number of competing definitions, applicable to differing physical systems, or a single definition as in Newtonian gravitation, or none at all. What happens in general relativity is that one is able to limit the range of possibilities to a considerable extent by means of various external criteria, but some ambiguity remains nonetheless.[15]

With these warnings in mind, let us press on to see how the apparatus of conformal completion functions in the $\Lambda \geq 0$ regime.

3. The Asymptotically Minkowski Case

Minkowski spacetime of course admits a conformal completion. Let us specialize to the four-dimensional case, where the Minkowski metric can be written in the form:

$$\frac{1}{(\cos\eta + \cos\chi)^2}\left(-d\eta^2 + d\chi^2 + \sin^2\chi\left(d\theta^2 + \sin^2\theta d\phi^2\right)\right),$$

[14] Geroch, "Asymptotic Structure," 3.   [15] Geroch, "Asymptotic Structure," 4.

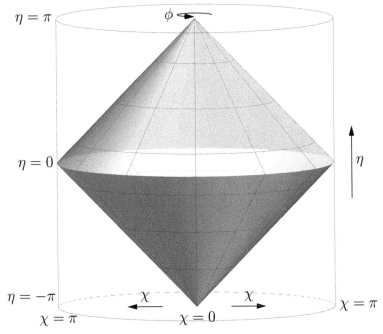

Figure 6.1 Minkowski spacetime is conformally equivalent to a region of the Einstein static universe. Fixing values of η and χ determines a sphere parameterized by ϕ and θ—with this sphere having maximum radius when $\eta = 0$ and degenerating to a point when $\eta = \pm\pi$. In the diagram, θ is suppressed, so that these spheres become circles. The conformal boundary of Minkowski spacetime corresponds to two portions of null cones—here the upper surface is $\eta + \chi = 0$ and the lower surface is $\eta - \chi = 0$. (Adapted, by permission of Cambridge University Press, from figure 3.3 of Griffiths and Podolský, *Exact Space-Times in Einstein's General Relativity*.)

with $\theta \in [0, \pi]$ and $\varphi \in [0, 2\pi)$ as usual and with with $\eta \in (-\pi, \pi)$ and $\chi \in [0, \pi]$, subject to the constraint that $-\pi < \eta + \chi < \pi$.

This tells us that Minkowski spacetime is conformally equivalent to a region of the Einstein static universe (see Figure 6.1). The conformal boundary of Minkowski spacetime comes in two pieces: $\mathcal{I}^+$, given by the conditions $\eta \in (0, \pi)$ and $\eta + \chi = \pi$; and $\mathcal{I}^-$, given by the conditions $\eta \in (-\pi, 0)$ and $\eta - \chi = -\pi$. Each has topology $\mathbb{R} \times S^2$. $\mathcal{I}^-$ and $\mathcal{I}^+$ each inherit a metric q_{ab} of degenerate signature from the embedding in the Einstein static universe and each also carries a distinguished null vector field n^a tangent to it (we can take $n^a = g^{ab} d\Omega$, where g^{ab} is the Einstein static metric and Ω is the conformal factor $\cos\eta + \cos\chi$). The vector field n^a induces a natural equivalence relation on $\mathcal{I}^\pm$ (with two points being equivalent if and only if they lie on an integral curve of n^a). The quotient of $\mathcal{I}^\pm$ by this equivalence relation has topology S^2 and inherits from q_{ab} the standard round metric. And, of course, if we find some other conformal transformation in virtue of which Minkowski spacetime admits a conformal completion, the relevant boundary will have geometric structures related to those just described by conformal transformations (in a certain sense).[16]

In rough analogy with our discussion of asymptotically anti-de Sitter spacetimes above, we introduce the following definitions. Let (M, g) be a four-dimensional spacetime that arises as a solution of some reasonable $\Lambda = 0$ Einstein-matter equations.

We say that (M, g) is *weakly asymptotically Minkowski* if it admits a conformal completion and its stress-energy tensor satisfies suitable fall-off conditions at conformal infinity.[17]

The conformal boundary of a weakly asymptotically Minkowski spacetime is a null manifold.[18] And assuming that (M, g) is weakly asymptotically Minkowski already has strong implications for the geometry of conformal infinity: every way of constructing a conformal infinity for (M, g) will endow it with a degenerate metric and a distinguished null vector field; and any two such representatives of the conformal infinity of (M, g) will be related to one another by a conformal transformation, in a suitable sense—indeed, the conformal boundaries of any two weakly asymptotically Minkowski spacetimes exhibit the same *local* conformal geometry.[19]

We say that a weakly asymptotically Minkowski spacetime is *strongly asymptotically Minkowski* if its conformal boundary has the same global structure as that of Minkowski spacetime: consisting of two disconnected components, each of topology $\mathbb{R} \times S^2$ and each of which is complete in a certain sense.[20]

---

[16] See, e.g., Geroch, "Asymptotic Structure," §3.

[17] For the relevant fall-off condition and further discussion, see, e.g., Ashtekar, "Geometry and Physics of Null Infinity."

[18] See Penrose, "Conformal Treatment of Infinity," lecture II.

[19] See Geroch, "Asymptotic Structure," §3.

[20] If (M, g) is weakly asymptotically Minkowski and has a conformal boundary of the prescribed topology, then we can always choose the conformal factor so that n^a is divergence-free. Relative to such a choice, we say that the conformal boundary is *complete* if the vector

The conformal boundary of a strongly asymptotically Minkowski spacetime has the same topology, local conformal geometry, and global structure as that of Minkowski spacetime.[21]

In the anti-de Sitter case considered above, we found that if we required a well-behaved $\Lambda < 0$ spacetime to have a conformal infinity with the same structure as that of anti-de Sitter spacetime, then this conformal boundary admitted a group of conformal isometries isomorphic to the isometry group of anti-de Sitter spacetime. By analogy, one might expect that the group of symmetries of the universal conformal boundary common to all strongly asymptotically Minkowski spacetime would be the Poincaré group. It is not. Rather, it is the Bondi-Metzner-Sachs (BMS) group, an infinite-dimensional extension of the Lorentz group. The 'extra' symmetries are so-called *super-translations* whose action on $\mathcal{I}^{\pm}$ can be pictured as follows: each component of $\mathcal{I}$ is ruled by a distinguished family of null geodesics (the integral curves of n^a); a supertranslation acts by translation along each of these geodesics, with the magnitude of the translation in general varying smoothly from one privileged null geodesic to another (see Figure 6.2).

The existence of supertranslations may seem like an unwelcome complication. But, in fact, it turns out that the BMS group has deep physical significance, one aspect of which is that the corresponding infinite-dimensional family of 'conserved' quantities can be used to describe the behaviour of gravitational waves far from sources—including the rate at which energy, momentum, and the like flow through $\mathcal{I}$.[22]

It is, of course, possible to impose stronger boundary conditions. In the $\Lambda < 0$ setting, requiring that a spacetime admit a conformal boundary with the same conformal structure as that of anti-de Sitter spacetime is equivalent to imposing so-called reflective boundary conditions—and implies that there can be no flux of energy or momentum through conformal infinity. As we have seen, in the $\Lambda = 0$ regime requiring that a spacetime admit a

---

field n^a is complete. For further discussion, see Ashtekar, "Geometry and Physics of Null Infinity," §2.

[21] See Wald, *General Relativity*, 278 ff., for a sense in which the geometry of a strongly asymptotically Minkowski spacetime is asymptotically flat at null infinity. Note that it is an open question whether being weakly asymptotically Minkowski already implies that $\mathcal{I}^{\pm}$ must have topology $\mathbb{R} \times S^2$. For results bearing on this question, see Newman, "The Global Structure of Simple Space-Times" (bearing in mind that the Poincaré conjecture has since been proved).

[22] In recent years, it has become clear that BMS group also plays a role in the memory effect and in describing the behaviour of soft gravitons. For an overview of these developments, see Ashtekar, Campiglia, and Laddha, "Null Infinity, the BMS Group, and Infrared Issues." For a detailed study of this and neighbouring terrain, see Strominger, *Lectures on the Infrared Structure of Gravity and Gauge Theory*.

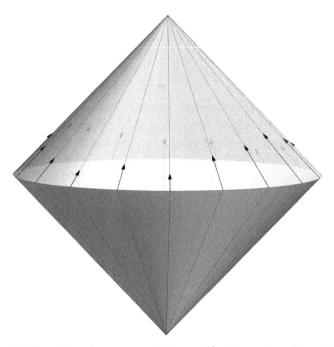

Figure 6.2 The action of a supertranslation on $\mathcal{I}^+$. The conformal boundary is ruled by null curves. A supertranslation is determined by choosing a time translation factor along each of these curves (which may vary smoothly from curve to curve).

conformal infinity with the same structure as that of Minkowski spacetime does not have this consequence. But it is still possible to further strengthen our boundary conditions in the $\Lambda = 0$ setting by imposing a version of reflective boundary conditions. And this would again have the consequence of ruling out the flux of conserved quantities through $\mathcal{I}$—and the further consequence that the group of symmetries that preserves these strengthened boundary conditions would be the good old Poincaré group.[23]

So the Poincaré group is *too small* to be useful as an asymptotic symmetry group if we are interested in modelling gravitational radiation. At the other extreme, the group of all diffeomorphisms of $\mathcal{I}^\pm$ would be *too large*— its structure is not rigid enough to determine an algebra of 'conserved' quantities that encode physical information. The BMS group is *just right* as

---

[23] For discussion, see Ashtekar, Bonga, and Kesavan, "Asymptotics with a Positive Cosmological Constant. I," §5.1.

an asymptotic symmetry group: with a structure rigid enough to determine an algebra of 'conserved' quantities that encodes physical information; large enough so that the flux of gravitational radiation at infinity can be modelled by these quantities.

QUESTION 6.2 (Larger). Some physicists maintain that a proper under-standing of the asymptotic symmetry group of general relativity suggests interesting morals concerning inter-theory relations.

> The enlargement of the asymptotic symmetry group from the Poincaré to the (infinite-dimensional) BMS group is an imprint left on the classical gravitational theory by the infrared behavior of the quantum gravitational field.[24]

> Since researchers were looking for symmetries that act in the asymptotic region where spacetime is almost flat, it was expected that one would reproduce the isometries of flat spacetime itself, namely the Poincaré group. Had this been the case, general relativity would reduce to special relativity at large distances and weak fields. Surprisingly... what they got instead was an infinite dimensional group, now called the BMS group. This contains as a subgroup the finite-dimensional Poincaré group. However, the four global translations are elevated to a whole function's worth of "supertranslations" that act independently on each point of the asymptotic sphere. Moreover, as we shall see, general relativity does not reduce to special relativity at large distances and weak fields. Instead, a large space of degenerate vacua remains.[25]

Is the picture that Ashtekar suggests here in tension with the idea that if we are to be able to apply our physical theories, lower-energy physics should not depend (much) on higher energy physics?[26] Are there other imprints of this kind of the quantum on the classical? What do we learn from them? Is the picture that Strominger suggests here consistent with the usual philosophers' ways of talking about reduction? What are the reasons for and against assimilating the phenomenon Strominger is discussing to more familiar examples of degenerate vacua in quantum field theory?

---

[24] Ashtekar, *Asymptotic Quantization*, xii (see also 100 f.).

[25] Strominger, *Lectures*, 74. For further discussion of asymptotic symmetries and degenerate vacua, see Hawking, Perry, and Strominger, "Soft Hair on Black Holes"; and Hawking, Perry, and Strominger, "Superrotation Charge and Supertranslation Hair on Black Holes."

[26] For sympathetic critical discussion of this idea, see Ruetsche, *The Physics of Ignorance*.

In these and other respects, are Ashtekar and Strominger right to think that the role of the BMS group in the analysis of gravitational radiation teaches us something interesting about the relation between general relativity and quantum gravity?

4. Going de Sitter

Let (M, g) be a four-dimensional spacetime that arises as a solution of some reasonable $\Lambda > 0$ Einstein-matter equations.

> We say that (M, g) is *weakly asymptotically de Sitter* if it admits a conformal completion and its stress-energy tensor satisfies suitable fall-off conditions at conformal infinity.[27]

The conformal boundary of any weakly asymptotically de Sitter spacetime is a spacelike manifold M equipped with a conformal equivalence class of Riemannian metrics.[28] Every compact manifold K ($\dim K \geq 3$ and odd) and every conformal equivalence class of Riemannian metrics on K arises as the conformal boundary of some weakly asymptotically de Sitter vacuum solution.[29]

> We say that a weakly asymptotically de Sitter spacetime is *strongly asymptotically de Sitter* if its conformal boundary has the same structure as that of de Sitter spacetime (conformally equivalent to a pair of round spheres).[30]

As we noted above, in the $\Lambda = 0$ setting, the notion of a strongly asymptotically Minkowski spacetime was admirably suited to the project of analyzing gravitational radiation. Does the notion of a (strongly or

---

[27] For the relevant fall-off condition and further discussion, see, e.g., Ashtekar, Bonga, and Kesavan, "Asymptotics. I," §2.

[28] See Penrose, "Conformal Treatment of Infinity," lecture II.

[29] On this point, see, e.g., Anderson, "Existence and Stability of Even-Dimensional Asymptotically de Sitter Spaces," theorem 1.1.

[30] Alternatively, we might require (M, g) to have a conformal infinity with the same structure as that of Schwarzschild-de Sitter spacetime or that of the cosmological patch of de Sitter spacetime—see, e.g., Ashtekar, Bonga, and Kesavan, "Asymptotics. I." For the sake of expository simplicity, we set these options aside, as their inclusion would not materially affect the points made below.

weakly) asymptotically de Sitter spacetime provide a similarly successful and flexible framework for understanding gravitational radiation in the $\Lambda > 0$ setting?

Unfortunately not.[31] Recall Geroch's warning from above: if your boundary conditions are too weak, interesting aspects of asymptotic behaviour will be invisible; if they are too strong, you will be excluding solutions that provide good models of the behaviour you are studying.

On the one hand, our notion of a weakly asymptotically de Sitter spacetime appears to be too weak to underwrite an analysis of gravitational radiation: even if the topology of the conformal boundary of spacetime is fixed, these boundary conditions place no further constraints on the (conformal) geometry of the boundary. In other words, the asymptotic symmetry group corresponding to these boundary conditions will include arbitrary diffeomorphisms of the boundary. This makes it impossible to single out one-parameter groups corresponding to translations or rotations. It would appear to be impossible to fashion sensible definitions of the flux of energy, momentum, or angular momentum through conformal infinity in this framework.

On the other hand, our notion of a strongly asymptotically de Sitter spacetime is too restrictive to underwrite a general-purpose analysis of gravitational radiation. Requiring that the conformal boundary of a spacetime have the same topology and conformal geometry as that of de Sitter spacetime has the salutary effect of determining an interesting group of asymptotic symmetries, isomorphic to the group of isometries of de Sitter spacetime. But this requirement turns out to be too strong, since it is equivalent to requiring that the magnetic portion of the asymptotic Weyl tensor vanishes, which implies that there can be no flux of 'conserved quantities' through $\mathcal{I}$ (in the vacuum regime).[32]

To summarize, if we do not strengthen boundary conditions we have no way of identifying quantities such as energy-momentum 'charges' and fluxes, needed to extract physics of the given isolated system. Alternatively, we can strengthen the boundary conditions and speak of de Sitter energy-momentum and angular momentum. But now these quantities cannot be radiated away, signalling that the restriction is unreasonably severe.[33]

---

[31] For an overview of the terrain merely sketched here, see Ashtekar, "Implications of a Positive Cosmological Constant for General Relativity"; for full details, see Ashtekar, Bonga, and Kesavan, "Asymptotics. I." For comparison with some other approaches, see Aneesh, Jahanur Hoque, and Virmani, "Conserved Charges in Asymptotically de Sitter Spacetimes."

[32] The parenthetical qualification is essential here—see the discussion of the Vaidya-de Sitter-de Sitter solution in Ashtekar, Bonga, and Kesavan, "Asymptotics. I."

[33] Ashtekar, "Implications," 4.

If the $\Lambda > 0$ regime supports reasonably general account of gravitational radiation to rival that of the $\Lambda = 0$ regime, then it must either involve boundary conditions at conformal infinity quite different from those considered above or depart from the strategy, so successful in the $\Lambda = 0$ regime, of conducting analysis at $\mathcal{I}$.[34]

A related problematic. In the $\Lambda = 0$ setting we have the following facts about mass.[35] Under modest assumptions, we have two important notions of mass for isolated systems. The Bondi mass lives at null infinity: it is defined by integrating over an 'instant of time at null infinity'—i.e., over a spherical spacelike cross-section of $\mathcal{I}$—a 'conserved' quantity associated with time translation. The ADM mass lives at spatial infinity: it is defined by integrating a conserved quantity associated with time translation over 'an instant of time at spatial infinity'—i.e., by taking the limit as $r \to \infty$ of integrals over spheres of radius r in a Cauchy surface. The ADM mass is genuinely conserved. And the celebrated positive mass theorem tells us that in all physically relevant cases, the ADM mass is positive (except for the case of Minkowski spacetime, in which it vanishes). The Bondi mass, on on the other hand, is typically *not* conserved—because gravitational radiation can leak through $\mathcal{I}$—but it too is non-negative in all cases of physical interest.

One would like to able to replicate at least some elements of this picture in the $\Lambda > 0$ regime. But, of course, already in de Sitter spacetime there are no globally defined time translations—which renders it impossible to give a general definition of the energy of a system located in a de Sitter spacetime. This problem carries over to the asymptotically de Sitter regime: time translations do not appear among the asymptotic symmetries according to any of the standard explications of the notion of an asymptotically de Sitter spacetime (this is intimately related to the fact that conformal infinity is spacelike). This leads to a number of open questions of longstanding. Is it possible to give a general definition of the mass for $\Lambda > 0$? Does some analog of the positive mass theorem hold? Can the standard framework of conserved quantities be generalized in some fruitful way to accommodate a positive cosmological constant?[36]

---

[34] For an approach of the first type, see Compère, Fiorucci, and Ruzziconi, "The Λ-BMS$_4$ Group of dS$_4$ and New Boundary Conditions for AdS$_4$." Ashtekar and co-workers have developed a framework of the second type: see Ashtekar, Bonga, and Kesavan, "Gravitational Waves from Isolated Systems"; Ashtekar, Bonga, and Kesavan, "Asymptotics with a Positive Cosmological Constant. I–III"; and Ashtekar and Bahrami, "Asymptotics with a Positive Cosmological Constant. IV."

[35] For discussion and references, see, e.g., Wald, *General Relativity*, §11.2.

[36] There is a large body of work on these and surrounding questions. Some good *entrées* into this literature: Kastor and Traschen, "A Positive Energy Theorem for Asymptotically de Sitter Spacetimes"; Kelly and Marolf, "Phase Spaces for Asymptotically de Sitter Cosmologies"; Penrose, "Cosmological Mass with Positive Λ"; and Ashtekar, "Implications."

Where does this discussion leave us? The notions of a weakly asymptotically de Sitter spacetime and of a strongly asymptotically de Sitter spacetime are clean and natural. In contrast to their $\Lambda = 0$ analogs, they prove to be unsuited for the analysis of conserved quantities and of the flow of gravitational radiation. But in the next chapter we will see that they have proved themselves useful for other purposes.

QUESTION 6.3 (Larger). Investigate the available strategies for analyzing gravitational radiation in the $\Lambda > 0$ context. Do the strengths and weaknesses of these teach us anything about philosophical topics such as approximation, idealization, and the physical content of boundary conditions?

7

Stability, Instability, and Hair

1. Prologue

In June 1918, Einstein conceded to Klein that de Sitter spacetime represents an empty world (see Section 2.1 above). Weyl held out until early the next year. In a paper submitted in October 1918 and published in January 1919, Weyl was still maintaining that Einstein's revised field equations admit no true vacuum solutions—and that de Sitter spacetime represents a world containing matter.[1] This paper caught Klein's attention and he wrote to Weyl to take up the issue, mentioning that Einstein had already run up the white flag.[2] Weyl replied with his own letter of surrender on 7 February 1919. Having conceded the chief point of contention, Weyl hastened, as Einstein had the previous summer, to emphasize to Klein that de Sitter spacetime is not static and to assert that it therefore cannot be taken seriously as a model of our world.[3] In this letter Weyl also makes an intriguing conjecture about the status of de Sitter spacetime within the family of $\Lambda > 0$ cosmological solutions: de Sitter spacetime is "separated by an abyss [*Abgrund*] from any static solution with the same topology [*Analysis-situs-Charakter*]."[4]

---

[1] See Weyl, "Über die statischen kugelsymmetrischen Lösungen von Einsteins ‚kosmologischen' Gravitationsgleichungen"; see also the discussion in §33 of the first edition of Weyl, *Raum-Zeit-Materie* (there is an expanded version of this discussion in §34 of Weyl, *Space-Time-Matter*). Weyl's view appears to be that, properly understood, the static patch of de Sitter spacetime models the gravitational field exterior to a singular matter distribution. Although Weyl would not have had this in mind, it might be helpful here to think of how the exterior Schwarzschild solution can be continued either by the interior Schwarzschild solution (representing a spherically symmetric massive body) or, less physically, by the Schwarzschild-Kruskal solution (with its two asymptotic ends, black hole, and white hole).

[2] For the relevant correspondence between Klein and Weyl, see Röhle, "Mathematische Probleme in der Einstein-de Sitter Kontroverse," Anhang B. The editors' notes to Einstein's concession postcard of 30 June 1918 (8.567) include portions of Weyl's letter of surrender (as well as a translation—followed below).

[3] Weyl tells Klein that in writing his letter he had consulted Einstein (who was visiting Zurich at the time) and that they are in agreement on the points just noted.

[4] It is not clear to me whether Weyl is presenting this, too, as a point on which he has Einstein's agreement.

Accelerating Expansion: Philosophy and Physics with a Positive Cosmological Constant. Gordon Belot, Oxford University Press. © Gordon Belot 2023. DOI: 10.1093/oso/9780192866462.003.0008

What does Weyl mean by this curious claim? One possibility is that he thinks that de Sitter spacetime is a true outlier: there just are no other vacuum solutions with spherical spatial sections. That may seem like a far-fetched reading. But in lecture notes of 1916–1917, Hilbert suggested that Minkowski spacetime was the only vacuum solution of the field equations of general relativity.[5] As late as 1932, the existence of further non-singular vacuum solutions was considered an open question.[6] And Choquet-Bruhat reports that in the early 1950s, Einstein still maintained that Minkowski spacetime was the only non-singular vacuum solution asymptotically flat at spatial infinity.[7] Perhaps once Weyl finally recognized that Einstein's amended field equations admit de Sitter spacetime as a $\Lambda > 0$ vacuum solution, he attributed to it the same sort of isolation that others had and would attribute to Minkowski spacetime in the $\Lambda = 0$ setting.

But here is another way of approaching the question. Already in 1918, Weyl was thinking about the initial value problem for the Einstein field equations.[8] He expects that suitable (local) Cauchy data will determine a unique (local) spacetime geometry. Let us fix a value $\Lambda_0 > 0$ for the cosmological constant. And let us consider the geometrically most natural time-slices of de Sitter spacetime and of the Einstein static universe: the spatial geometry is homogeneous and isotropic, the matter distribution is homogeneous, and both matter and geometry are invariant under time-reversal (so we don't have to worry about rates of change). In the de Sitter case, these spatial slices correspond to round three-spheres of radius $r_0 := \sqrt{3/\Lambda_0}$ with matter density $\rho_0 \equiv 0$. In the Einstein static universe, they correspond to round three-spheres of radius $r_1 = \sqrt{1/\Lambda_0}$ with matter density $\rho_1 \equiv \Lambda_0/4\pi$. Obviously, we can also consider initial data sets that interpolate between these two cases.

---

[5] See the passages from Hilbert's lecture notes quoted by Renn and Stachel, "Hilbert's Foundations of Physics," p. 947. See also Hilbert, "The Foundations of Physics (Second Communication)," 63–66. This may seem strange, since the partial differential equations arising in physics typically have infinite-dimensional families of solutions—but for a famous example of a partial differential equation with exactly one solution, see, e.g., John, *Partial Differential Equations*, 235.

[6] See Lanczos, "On the Problem of Regular Solutions of Einstein's Gravitational Equations." In earlier papers Lanczos had actually claimed that there could be no such solutions—for discussion and references, see Stachel, "Lanczos's Early Contributions to Relativity and His Relationship with Einstein," 202 f.

[7] Choquet-Bruhat, *A Lady Mathematician in this Strange Universe*, 118. Pauli, writing in 1958, regarded this as an open question—see *Theory of Relativity*, Supplementary Note 17.

[8] See §33 of the first edition of Weyl, *Raum-Zeit-Materie*; or the expanded version of this discussion in §34 of Weyl, *Space-Time-Matter*.

I suggest that part of what Weyl intended when he was saying that there is an abyss between de Sitter spacetime and any static solution with the same topology is the following:

> *Abgrund* Conjecture. Data sets intermediate between de Sitter data and Einstein static data will evolve to yield solutions that are not much at all like de Sitter spacetime—but if they are similar enough to Einstein static data, then they will evolve to solutions pretty similar to the Einstein static universe.

If this reading of Weyl is correct, then he was quite mistaken—and also very close to being right. For, as we will see below, the *Abgrund* Conjecture is essentially correct *if you reverse in it the roles played by de Sitter spacetime and by the Einstein static universe.* The main point of this chapter is to explain that observation and to provide some context for thinking about its significance.

We will begin in Section 2 with a review of some fairly precise notions of stability from the theory of dynamical systems (= systems with finitely many degrees of freedom, described by ordinary differential equations). In Section 3, we introduce the related (but less precise) notions that will be our main focus here: the notions of global non-linear stability and instability of particular solutions of partial differential equations. In the Sections 4, 5, 6, and 8, we survey how these notions apply to some central solutions of the Einstein field equations: the Einstein static universe (unstable); Minkowski spacetime (stable); de Sitter spacetime (stable); anti-de Sitter spacetime (unstable under reflective boundary conditions, conjectured to be stable under dissipative boundary conditions). In Section 7, we discuss a sense in which de Sitter spacetime is conjectured to be a dynamical attractor in a sense yet stronger than global non-linear stability.

QUESTION 7.1 (Smaller?). In a passage added to §34 of the fourth edition of *Raum-Zeit-Materie* in 1921 (translated into English as *Space-Time-Matter*) Weyl offers two versions of de Sitter geometry: Klein's preferred hyperboloid representation and the static patch. He remarks:

> The question arises whether it is the first or the second co-ordinate system that serves to represent the whole world in a regular manner.

He then goes on to say on de Sitter's view, the first coordinate system accurately represents the world, which is void and non-static, while on Einstein's view, it is the second coordinate system that accurately represents a static world containing matter. This way of characterizing the dispute can be found already in Weyl's letter to Klein of 20 September 1918, where Weyl further suggests that the question at hand is one for physics rather than mathematics. So did his exchange with Klein in early 1919 really change Weyl's mind about anything?[9]

2. Warmup: Stability and Instability in Dynamical Systems

Consider three sets of ordinary differential equations, each describing trajectories $\gamma(t) = (x(t), y(t))$ in the xy-plane.[10]

(1) $\dot{x} = \frac{1}{2}x - y - \frac{1}{2}(x^3 + y^2 x)$ $\qquad \dot{y} = x + \frac{1}{2}y - \frac{1}{2}(y^3 + x^2 y)$
(2) $\dot{x} = y$ $\qquad \dot{y} = -x$
(3) $\dot{x} = -x - y - x(x^2 + y^2)$ $\qquad \dot{y} = x - y - y(x^2 + y^2)$

Each of these equations admits the trivial solution $\gamma_0(t) \equiv (0,0)$. Away from the origin, we can recast our equations in terms of polar coordinates. We then have:

(1*) $\dot{\theta} = 1$ $\qquad \dot{r} = r(1 - r^2)/2$
(2*) $\dot{\theta} = 1$ $\qquad \dot{r} = 0$
(3*) $\dot{\theta} = 1$ $\qquad \dot{r} = -r - r^3$

Since in each case $\dot{\theta} \neq 0$, the trivial solution is the only time-independent solution admitted by any of our three systems of equations. We are interested in the behaviour of solutions of these systems with initial data near the origin (see Figure 7.1). For the first set of equations, we see that if $0 < r < 1$, then $\dot{r} > 0$. So solutions that start anywhere near the origin spiral outwards—there is no way to constrain a non-trivial solution to remain as close as we like to the origin. Contrast this with the behaviour determined by the second set of equations: solutions correspond to circles concentric with the origin—so we can constrain a solution to permanently remain as close as we like to the

[9] The relevant letters can be found in Röhle, "Mathematische Probleme in der Einstein-de Sitter Kontroverse," Anhang B. In this connection, see also Einstein's postcard of surrender to Klein of 20 June 1918 (8.567).
[10] For these examples, see Hirsch, Smale, and Devaney, *Differential Equations, Dynamical Systems, and an Introduction to Chaos*, chapters 2 and 8.

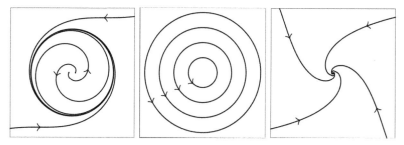

Figure 7.1 On the left, the behaviour of solutions of System (1): initial data near the origin determine solutions that spiral away from the origin. In the centre, the behaviour of solutions of System (2). On the right, the behaviour of solutions of System (3): arbitrary initial data determine solutions that spiral inwards, asymptotically approaching the origin.

origin by constraining its initial data to lie sufficiently close to the origin. This is also possible with solutions of the third set of equations. Indeed, something stronger obtains in this case: away from the origin, $\dot{r}$ is always negative, so solutions that begin anywhere in the plane spiral inwards—indeed, each such solution asymptotically approaches the origin as $t \to \infty$.

In general, in the setting of ordinary differential equations (with independent variable t), a solution $f(t)$ is considered an *equilibrium solution* if it is t-independent. An equilibrium solution $f(t) \equiv X$ is called *stable* if each open neighbourhood U_0 of X in the space of initial data contains an open neighbourhood U of X such that solutions determined by initial data in U remain permanently in U_0—and it is considered *asymptotically stable* if, further, U can be chosen so that solutions determined by initial data in U tend towards X in the $t \to \infty$ limit.[11] Note that in this setting, the space parameterized by the dependent variables of a differential equation (x and y in our examples) will be a finite-dimensional manifold, and so come equipped with a natural topology.

It is customary to dismiss unstable equilibria as being useless from a practical point of view. Here is a discussion from a classic textbook on ordinary differential equations and dynamical systems:

The study of equilibria plays a central role in ordinary differential equations and their applications. An equilibrium point, however, must satisfy a

---

[11] Note that *asymptotic stability* is stronger than *stability*—in contrast with *asymptotically de Sitter*, which is weaker than *de Sitter*.

certain stability criterion to be significant physically. (Here, as in several other places in this book, we use the word *physical* in a broad sense; in some contexts, physical could be replaced by *biological, chemical,* or even *economic.*)

An equilibrium is said to be *stable* if nearby solutions stay nearby for all future time. In applications of dynamical systems one cannot usually pinpoint positions exactly, but only approximately, so an equilibrium must be stable to be physically meaningful.[12]

The intuition here is straightforward: a rigid pendulum (a weight on a rod, attached to a pivot, free to rotate in a vertical plane) has two equilibrium states—one in which the bob is at rest at the lowest possible point, the other in which it is at rest at the highest possible point. We might well find a pendulum that is persistently in the first sort of state, so far as we can tell— but we don't expect to ever find one that is, so far as we can tell, persistently in the second sort of state. And this is because resting at the lowest point is a stable equilibrium for this system, while resting at the highest point is an unstable equilibrium.[13] Measurements of finite accuracy could tell us that the pendulum was at least approximately in one of these two equilibrium states— the dynamics then tells us that we should expect this situation to persist if and only if the equilibrium in question is the stable one (if we are half-decent at measuring). So the stable equilibrium has a sort of usefulness for modelling and prediction that the unstable equilibrium lacks.[14]

QUESTION 7.2 (Larger?). A similar picture is prevalent in discussion of solutions of the Einstein equations. Examples:

A physical theory is a correspondence between certain physical observa-tions and a mathematical model (in this case a manifold with Lorentz

---

[12] Hirsch, Smale, and Devaney, *Differential Equations*, 174. For more nuanced philosophical discussions of the role of stability in accounts of explanation and causation, see Batterman, *The Devil in the Details*; and Woodward, "Sensitive and Insensitive Causation."

[13] Amazingly, the latter becomes stable if we rapidly jiggle the pivot point up and down in the right way—see Arnold, *Mathematical Understanding of Nature*, chapter 15.

[14] Or, better, that is so except perhaps in special circumstances. We need to guard against over-statement: one can picture situations—involving short timescales of interest or a lot of friction at the pivot point—in which even the unstable equilibrium might provide acceptable predictions; and likewise for the unstable equilibrium of the first set of equations considered above—if the timescale on which r changes is very large compared to our measurement timescale, then even the unstable equilibrium state might provide a decent model of some situations.

metric). The accuracy of the observations is always limited by practical difficulties and by the uncertainty principle. Thus the only properties of space-time that are physically significant are those that are stable in some appropriate topology.[15]

One further area in which global methods [in general relativity] may prove useful is in questions concerning stability. It is a general feature of the description of physical systems by mathematics that only conclusions which are stable, in an appropriate sense, are of physical interest.... Suppose for example that we give initial data for a system and from this predict its evolution. In any actual situation, of course, we could only determine the actual initial data up to some specified error. But if a prediction depends crucially upon the precise data—if it undergoes a drastic change under even arbitrarily small perturbations of that data—then our prediction, while perhaps suggestive and useful, has little physical significance.[16]

If an explicit solution of the Einstein equations is to be relevant to the description of reality, then it must have some stability properties.[17]

Are the considerations these authors are adverting to epistemological, methodological, or metaphysical? What are the salient differences, if any, between the way such considerations function in thinking about general relativity and in thinking about other theories?[18]

3. Global Stability

A variety of related notions of stability and instability are widely applied to systems of partial differential equations. For present purposes, the most salient is the notion of *global stability*: schematically, a solution ϕ_0 of a system of equations Δ is *globally stable* if initial data sets sufficiently close to initial data sets determined by ϕ_0 determine (maximal) solutions of Δ that are globally similar to ϕ_0. A few clarifications.

---

[15] Hawking, "Stable and Generic Properties in General Relativity," 395.
[16] Geroch, "General Relativity in the Large," 70.
[17] Rendall, *Partial Differential Equations in General Relativity*, 193.
[18] For extant philosophical discussions of this territory, see Fletcher, "Similarity, Topology, and Physical Significance in Relativity Theory"; Fletcher, "The Principle of Stability"; and Wu, "Stability, Genericity, and 'Physicalness' in Spacetimes."

a) Obviously this is only a schema. In concrete applications, we need to say what we mean by *initial data sets* (in general relativity, we often— but not always—mean *initial data posed on Cauchy surfaces*), we need to specify a topology or norm on the space of initial data that allows us to make sense of *sufficiently similar*, and we need to stipulate the relevant sense of *global similarity*.

b) When considering non-linear differential equations, we will follow custom and speak of *global non-linear stability* to emphasize that we are concerned with the (in)stability of solutions relative to the full non-linear equations rather than with respect to the linearized approximation to these equations.[19]

c) It is important to distinguish the notion of global (non-linear) stability from another important notion of stability. Schematically, one says that a differential equation has a *well-posed initial value problem* if: (i) each admissible initial data set determines exactly one solution; and (ii) solutions depend continuously on initial data (where this is to be understood locally in time, rather than globally).[20] The standard name for the second of these requirements is *Cauchy stability*. Cauchy stability is a property of a system of differential equations, while global stability is a property of individual solutions of such equations. As we will see below, some highly natural solutions of the Einstein equations fail to enjoy global non-linear stability—even while the Einstein equations themselves exhibit Cauchy stability.[21]

---

[19] If ϕ is a well-behaved solution of some non-linear equations Δ, then it often makes physical sense to consider the linear equations that result from 'linearizing around ϕ' (i.e., doing a power series expansion and dropping terms of higher than linear order). Heuristically, a solution $\delta\phi$ to the linearized equations at ϕ can be thought of as an element of the tangent space $T_\phi S$ at ϕ to the space of solutions S of the full non-linear equations Δ. We say that Δ enjoys *linearization stability* at ϕ if this heuristic picture can be substantiated: that is, if for every $\delta\phi$ there is a one-parameter family of solutions $\phi(t)$ to Δ such that $\phi(0) = \phi$ and $\dot\phi(0) = \delta\phi$. The Einstein equations enjoy linearization stability at some but not all solutions commonly used in applications—see Girbau and Bruna, *Stability by Linearization of Einstein's Field Equation*. On the pitfalls of relying on linearizations in studying non-linear equations, see Ringström, *On the Topology and Future Stability of the Universe*, §2.4.

[20] In order be rendered precise in application to a system of partial differential equations, this schema needs to be filled out with a choice of functions spaces (for initial data and for solutions) equipped with topologies. For helpful discussions and examples, see Egorov and Shubin, *Foundations of the Classical Theory of Partial Differential Equations*, §1.8. For an illuminating example in which solutions fail to depend continuously on initial data, see Ringström, *Topology and Future Stability*, §2.3.

[21] On the Cauchy stability of general relativity, see Hawking and Ellis, *The Large Scale Structure of Space-Time*, §§7.6 f.; Wald, *General Relativity*, §10.2; and Ringström, *The Cauchy Problem in General Relativity*, chapter 15.

Let us turn to some examples. In each case, we will be interested in the question of the global (in)stability of the trivial solution $u_0(t,x) \equiv 0$ of the equation in question.

i) Consider, first, a nice example—the wave equation in flat spacetime:

$$\partial_{tt} u(t,x) - \partial_{xx} u(t,x) = 0,$$

with $t \in \mathbb{R}$ and $x \in \mathbb{R}^3$. For this equation, specifying an initial data set amounts to specifying behaviour of the field u and its first time derivative at an initial time. In this case, there are reasonable senses in which the trivial solution u_0 is globally stable: e.g., compactly supported smooth initial data sets determine solutions that decay to the trivial solution as $t \to \pm\infty$.[22]

ii) Consider the heat equation for a one-dimensional bar of unit length:

$$\partial_t u(t,x) - \partial_{xx} u(t,x) = 0,$$

with $t \in [0,\infty)$ and $x \in [0,1]$. In this case, one specifies an initial data set by specifying the behaviour of the field at $t=0$. It turns out that the trivial solution can be either globally stable or globally unstable, depending on the boundary conditions imposed on the ends of the bar.

A natural choice is to pose Dirichlet boundary conditions at both ends, requiring that $u(t,0) = 0$ and $u(t,1) = 0$ for all $t \geq 0$. Physically, this means that some external agency keeps the ends of the bar at temperature 0—which is consistent with heat flowing out through the ends of the bar. In this case, the solution determined by any continuous initial data set will decay rapidly towards u_0 in the $t \to \infty$ limit.[23] So it is natural to consider the trivial solution to be globally stable—arbitrary continuous perturbations of trivial initial data lead to solutions that are very similar indeed to the trivial solution.

But things look very different if we instead impose weirdo hybrid Dirichlet-Robin boundary conditions, requiring that for all $t \geq 0$ we have $u(t,0) = 0$ and $\partial_x u(t,1) - 2u(t,1) = 0$ (these model a situation where the left end of the bar is held at a constant temperature while the

---

[22] For a precise statement, see Sogge, *Lectures on Non-Linear Wave Equations*, theorem I.1.1.
[23] See, e.g., Olver, *Introduction to Partial Differential Equations*, §4.1.

right end is attached to a heat source that pumps more heat into the right end of the bar the hotter it is). With these boundary conditions the heat equation has runaway solutions: the right end of the bar gets hotter and hotter, so more and more heat is pumped into the bar—and as a result, the temperature diverges exponentially as a function of time at every point of the bar other than its left endpoint.[24] For any reasonable way of measuring such things: arbitrarily close to the trivial initial data, one can find initial data that determine such runaway solutions. Since the resulting solutions differ dramatically from the trivial solution in the $t \to \infty$ limit, it is natural to consider this a case where the trivial solution $u_0(t, x) \equiv 0$ globally unstable.

iii) Consider, finally, another nasty example—the non-linear wave equation

$$\partial_{tt} u - \partial_{xx} u - u^2 = 0$$

($t \in [0, \infty)$ and $x \in \mathbb{R}^3$). An initial data set again consists of a specification of the behaviour of the field and of its first time-derivative at $t = 0$. Any initial data set according to which the field and its time-derivative vanish outside of a compact set K and have positive integrals over K determines a solution that blows up in finite time (and so fails to exist globally).[25] So, in particular, there are initial data sets arbitrarily close to the trivial initial data set determined by $u_0(t, x) \equiv 0$ that determine solutions that are singular, and hence globally very different from the trivial solution u_0. It is natural to consider this an instance of global non-linear instability.

QUESTION 7.3 (Smaller?). Marsden and Ratiu reach the following conclusion about a certain sort of equilibrium solution of a set of equations from fluid mechanics:

we have stability with respect to one distance measure, but not with respect to another. The distinction is reflecting important physical mechanisms.[26]

Can you find examples with the same feature in general relativity? What lessons can be learned from such examples?

---

[24] For details see, e.g., Olver, *Introduction*, 134 ff.
[25] See, e.g., Evans, *Partial Differential Equations*, §12.5.2.
[26] Marsden and Ratiu, "Nonlinear Stability in Fluids and Plasmas," 122.

QUESTION 7.4 (Larger?). Part of the art of establishing Cauchy stability for a partial differential equation lies in selecting the right topologies for the relevant spaces of initial data and solutions—in the case of equations like the vacuum Einstein equations, for example, Cauchy stability holds if we use suitable Sobolev spaces, but fails for some other natural function spaces.[27] Are choices that make our theories well-behaved in this particular sense *better* in some way than ones that don't? How? Why? Is there some requirement to think of theories as encoding a continuous dependence of histories on instantaneous states? If so, what sort of requirement is it? What sort of mistake am I making if I insist on using the 'wrong' topology?

QUESTION 7.5 (Larger?). Are there interesting physical theories that fail to exhibit Cauchy stability? How do such theories mesh with philosophical accounts of causation on which in order for actual event C to cause actual event E it is necessary that if an event sufficiently similar to C had occurred, an event sufficiently similar to E would have occurred?[28]

QUESTION 7.6 (Smaller? Larger?). In fields such as geophysics and medical imaging one is often concerned with inverse problems: the problem of using measurements made in the region exterior to an object to confirm or to disconfirm hypotheses concerning that object's internal structure. Some inverse problems are well-posed: one can choose interesting function spaces parameterizing the possible internal structures of the object and the possible outcomes of measurements that can be made in the exterior region in such a way that there is one and only one internal structure consistent with any complete set of external measurements (existence and uniqueness) and such that this unique internal structure depends continuously on the exterior measurements that determine it (stability). It is a basic fact of life that many inverse problems of practical interest are ill-posed due to instability.[29] By analogy with the sort of considerations adduced above in favour of the crucial role of stability considerations in physics (the need to be able to draw reliable inferences from noisy data), one might expect such problems to be considered intractable. But there are methods (such as Tikhonov

---

[27] See Ringström, *Topology and Future Stability*, §§2.2 f.
[28] For an account of this kind, see Lewis, "Causation as Influence." For further discussion of such theories, see Woodward, "Sensitive and Insensitive Causation."
[29] For examples, see, e.g., Isakov, *Inverse Problems for Partial Differential Equations*; or Natterer, *The Mathematics of Computerized Tomography*.

regularization) that allow one to work around such instability.[30] Does this suggest that the case made above for the centrality of stability considerations may have been over-stated in some respects?[31]

4. Example: The Instability of the Einstein Static Universe

Let us finally move on to our main target: senses in which various natural solutions of the Einstein equations are or are not globally non-linearly stable.

Consider, first, the Einstein static universe, which is both static and spatially homogeneous. In 1917 (and for a number of years afterwards) Einstein advocated it as a good model of our Universe. Einstein of course recognized that this could be true only relative to a certain level of idealization.

> If I imagine the universe divided into regions of equal size, each of which contains on average 1,000 fixed stars, then each of these regions will contain approximately the same amount of mass. I.e., I make the hypothesis that apart from the local concentration in stars, the matter is uniformly distributed on the large scale. This matter I replace, for the sake of convenience, with homogeneously distributed matter of the same mean density. In this way, the grav. field's local structure is admittedly changed, or disturbed compared to reality. However, the metric character of this field will be preserved *on the large scale*, so I am correctly informed about the geometric nature of the universe on the large scale. Thus, in this consideration I abstract totally from the structure of the field in spaces of the order of magnitude or smaller than the distance from neighboring fixed stars.[32]

---

[30] Tikhonov regularization plays a central role not only in the mathematics of inverse problems exhibiting instability (see the references of the preceding note), but also in machine learning and in statistical learning theory—see, e.g., Shalev-Shwartz and Ben-David, *Understanding Machine Learning*, chapter 13; and Hastie, Tibshirani, and Friedman, *The Elements of Statistical Learning*, chapter 5.

[31] On this topic, see Gyenis, *Well Posedness and Physical Possibility*.

[32] From Einstein's letter to Mie of 22 February 1918 (8.470). See also his potato letter to de Sitter of 22 June 1917 (8.356).

During this period, Einstein was also convinced that the structure of the Universe was time-independent on a cosmological scale.[33] So it seems like he was committed to maintaining that his static universe was globally non-linearly stable in the sense that Cauchy initial data sets close to Cauchy initial data from the static universe should evolve to spacetimes that were globally similar to the static universe in being at least approximately static in some reasonable sense.

In 1930, Eddington showed that Einstein's static universe is not stable in this sense. Einstein managed to construct a static solution to the $\lambda > 0$ Einstein-dust equations by precisely balancing the density of matter (with its tendency to foment collapse) against the magnitude of the cosmological constant (with its tendency to drive expansion).[34] Eddington restricts attention to spatially isotropic solutions of the Einstein-dust equations with cosmological constant $\lambda > 0$ and spatial topology S^3. Each such solution is characterized by two functions of time t: $a(t)$, describing the volume of space at time t; and $\rho(t)$, describing the spatially homogeneous density of matter at time t. These are related by the equation:

$$3\frac{d^2}{dt^2}a(t) = a(t) \cdot (\lambda - 4\pi\rho(t)).$$

So, Eddington observes,

For equilibrium (Einstein's solution) we must accordingly have $\rho = \lambda/4\pi$. If now there is a slight disturbance so that $\rho < \lambda/4\pi$, da^2/dt^2 is positive and the universe accordingly expands. The expansion will decrease the density; the deficit thus becomes worse, and da^2/dt^2 increases. Similarly if there is a slight excess of mass a contraction occurs which continually increases. Evidently Einstein's world is unstable.[35]

---

[33] He was confident enough in this claim that he dismissed from consideration de Sitter spacetime, on the grounds that it does not admit a global static frame. See the letter to Klein of before 3 June 1918 (8.556), the letter to de Sitter of 14 June 1917 (8.351), the postcard to Klein of 20 June 1918 (8.567), and the postcard to Weyl of 23 May 1923 (14.40). For further discussion of the strong sense in which Einstein took our universe to be static, see Belot, *Cosmological Reconsiderations*, chapter IV.

[34] Eddington, "On the Instability of Einstein's Spherical World." As Eddington notes, aspects of his analysis are prefigured in Lemaître, "A Homogeneous Universe of Constant Mass and Increasing Radius Accounting for the Radial Velocity of Extra-Galactic Nebulae." For helpful context and discussion, see Luminet, "Editorial Note to: Georges Lemaître, A Homogeneous Universe of Constant Mass and Increasing Radius Accounting for the Radial Velocity of Extra-Galactic Nebulae."

[35] Eddington, "Instability," 670.

An *Eddington-Lemaître solution* is spatially isotropic but slightly under-dense at $t = 0$ (relative to the Einstein static universe). Such solutions are asymptotic to the Einstein static universe in the past and to de Sitter space-time in the future (or vice versa)—and hence are strongly asymptotically de Sitter in one temporal direction while not admitting a conformal completion in the other.[36] Since initial data arbitrarily close to Einstein static initial data can determine Eddington-Lemaître solutions, it is natural to regard the Einstein static universe as being globally non-linearly unstable.

Even local over- or under-densities can be expected to lead to runaway contraction or expansion.[37] So Einstein's static universe is more or less useless for its intended role as a model of our world on the largest scale—generically, in a universe in which the distribution of matter exhibits even the slightest inhomogeneity, the Einstein static universe will present a radically misleading picture of the past or future.[38]

5. The Global Non-Linear Stability of Minkowski Spacetime

A celebrated theorem of Christodoulou and Klainerman establishes that Minkowski spacetime is globally non-linearly stable in an important sense: in the $\Lambda = 0$ regime, vacuum Cauchy initial data sets sufficiently close to the trivial initial data that the Minkowski metric induces on flat hypersurfaces determine inextendible vacuum solutions that are globally similar to Minkowski spacetime—in particular, such solutions are geodesically complete and there is a sense in which their metrics asymptotically approach the Minkowski metric as one proceeds to infinity along any geodesic.[39] A couple of points to note.

[36] According to our definition, if an n-dimensional Lorentz manifold has a conformal completion, then the corresponding conformal boundary must be $(n-1)$-dimensional. The Einstein static universe doesn't admit a conformal completion in this sense: any conformal factor that renders the temporal direction finite also pinches spatial sections down to a point—see Penrose, "Conformal Treatment of Infinity," lecture III.

[37] It is known, however, that the Einstein static universe is stable against certain perturbations—see, e.g., Barrow *et al.*, "On the Stability of the Einstein Static Universe."

[38] For another example of the instability of a class of highly symmetric cosmological models, see Ringström, "Instability of Spatially Homogeneous Solutions in the Class of T^2-Symmetric Solutions to Einstein's Vacuum Equations"; or LeFloch and Smulevici, "Future Asymptotics and Geodesic Completeness of Polarized T^2-Symmetric Spacetimes."

[39] For an accessible overview, see Rendall, *Partial Differential Equations*, §10.3. For one further level of detail, Christodoulou, *Mathematical Problems of General Relativity*, chapter 4. Full details can be found in Christodoulou and Klainerman, *The Global Nonlinear Stability of the Minkowski Space*.

a) This is a *small-data* result—all bets are off for initial data that are not sufficiently close to the trivial Minkowski initial data. Indeed, there are initial data defined on $\mathbb{R}^3$ that determine $\Lambda = 0$ vacuum spacetimes that fail to be globally similar to Minkowski spacetime in the relevant sense—e.g., because they include singularities or because they feature phenomena such as plane gravitational waves that mean that there are certain directions in which the spacetime geometry does not approach Minkowski geometry, no matter how far out towards infinity one proceeds.[40]

b) One might have hoped that small perturbation of Minkowski initial data would determine solutions that admitted conformal completions with the same global structure as that of Minkowski spacetime. The Christodoulou-Klainerman result does *not* deliver this conclusion. Nor do any of the subsequent generalizations of this result.[41] Indeed, it now appears that only fairly special small perturbations of Minkowski initial data have this feature—the problem being that the existence of a well-behaved conformal completion implies fall-off conditions that are somewhat too strong to be satisfied by generic perturbations of Minkowski spacetime.[42]

6. The Global Non-Linear Stability of de Sitter Spacetime

There are multiple senses in which de Sitter spacetime is globally non-linearly stable within the vacuum $\Lambda > 0$ realm.

---

[40] For plane gravitational waves, see, e.g., Griffiths and Podolský, *Exact Space-Times in Einstein's General Relativity*, §17.5. Note that according to another celebrated theorem of Christodoulou, black holes can form via the focusing of weak gravitational waves: there exist vacuum $\Lambda = 0$ initial data sets defined on $\mathbb{R}^3$ that correspond to the presence of widely dispersed and weak gravitational waves, and which determine solutions featuring black holes—for an overview, see Dafermos, "The Formation of Black Holes in General Relativity (after D. Christodoulou)."

[41] On this point, see Bieri, "An Extension of the Stability Theorem of the Minkowski Space in General Relativity," §1; Klainerman and Nicolò, *The Evolution Problem in General Relativity*, 357; and Lindblad and Rodianski, "The Global Stability of Minkowski Space-Time in Harmonic Gauge," 1406 f.

[42] For discussion and references, see Valiente Kroon, *Conformal Methods in General Relativity*, chapters 12 and 18; Friedrich, "Geometric Asymptotics and Beyond," §4; and Kehrberger, "The Case Against Smooth Null Infinity." Note that there are other ways of making sense of the notion of null infinity, better suited to the framework used in analysis of the global non-linear stability of Minkowski spacetime—on this point, see Klainerman and Nicolò, *The Evolution Problem*, chapter 8.

Friedrich has shown that if (M, g) is a four-dimensional weakly asymptotically de Sitter vacuum spacetime with a conformal boundary given by two copies of a compact manifold K, then any vacuum Cauchy data sufficiently close to Cauchy data from (M, g) determines a maximal globally hyperbolic solution that is null geodesically complete and weakly asymptotically de Sitter with conformal boundary given by two copies K.[43] So de Sitter spacetime itself is non-linearly stable against vacuum perturbations. In particular, the existence and topology (although not, in general, the geometry) of the conformal boundary of de Sitter spacetime is stable under such perturbations.[44] So here the picture of the preceding chapter is inverted and we have a sense in which the apparatus of conformal completion functions *better* in the $\Lambda > 0$ setting than it does in the $\Lambda = 0$ setting.

A result due to Ringström establishes a weaker form of future non-linear stability for a wide family of solutions of cosmological interest. Let (M, g) be a solution of the $\Lambda > 0$ vacuum field equations.[45] Suppose, further, that (M, g) is future geodesically complete and admits a slicing into locally homogeneous Cauchy surfaces (satisfying certain further technical restrictions). Then Cauchy data sufficiently similar to locally homogeneous Cauchy data for (M, g) evolve to yield solutions that are likewise future geodesically complete, and which locally satisfy certain asymptotic expansions that give some control over the sense in which the perturbed solution is asymptotic to the unperturbed solution.[46]

---

[43] For a review of the relevant results, see Friedrich, "Geometric Asymptotics and Beyond," §3; and Valiente Kroon, *Conformal Methods*, chapter 15. Note that the result holds also for certain types of matter. For a discussion of completeness and an extension of these results to higher even dimensions, see Anderson, "Existence and Stability of Even-Dimensional Asymptotically de Sitter Spacetimes." Note, further, that in four spacetime dimensions, the assumption that (M, g) has well-behaved conformal boundaries both to the past and the future in fact imposes strong constraints on the topology of M—see Anderson, "On the Structure of Asymptotically de Sitter and Anti-de Sitter Spaces," 872.

[44] For discussion, see Ashtekar, Bonga, and Kesavan, "Asymptotics with a Positive Cosmological Constant. I," §4.2.

[45] Or, more generally, of a set of Einstein-non-linear scalar field equations featuring a potential term that causes the scalar field to mimic a positive cosmological constant.

[46] For an overview, see Ringström, "On Proving Future Stability of Cosmological Solutions with Accelerated Expansion," §3; for details, Ringström, "Future Stability of the Einstein-Non-Linear Scalar Field System." For related results for the Einstein-Vlasov equations, see Ringström, *Topology and Future Stability*. For a result concerning global non-linear stability of slowly rotating Kerr-de Sitter solutions, see Hintz and Vasy, "The Global Non-Linear Stability of the Kerr-de Sitter Family of Black Holes."

7. The Cosmic No-Hair Conjecture

Here is a *very* rough and intuitive picture. In an expanding $\Lambda > 0$ universe, the tendency towards expansion due to Λ is undiluted by expansion even while expansion tends to dilute the efficacy of other forms of energy. So almost no matter how the universe starts, the cosmological constant should become the main driver of cosmic evolution—so that eventually the universe should end up looking more and more like de Sitter spacetime. So de Sitter spacetime is a dynamical attractor in a strong sense that goes well beyond the sort of global non-linear stability results canvassed in the preceding section.

The *cosmic no hair conjecture* adds some nuance to this intuitive picture, while holding on to the core idea that the repulsive dynamics driven by a positive cosmological constant should have the effect of isotropizing instantaneous physical states at late times. A rough statement of the conjecture as it is currently understood goes something like:

COSMIC NO-HAIR CONJECTURE. For reasonable matter sources, generic future geodesically complete $\Lambda > 0$ solutions should look like de Sitter spacetime to late-time observers—i.e., generic such solutions contain Cauchy surfaces Σ such that for any inextendible timelike curve γ, $J^+(\Sigma) \cap J^-(\gamma)$ can be sliced into spacelike surfaces relative to which it becomes arbitrarily similar at late times to the cosmological patch of de Sitter spacetime relative to its spatially flat slicing.[47]

A couple of remarks.

i) The restriction to geodesically complete solutions is required to rule out black hole spacetimes—one certainly doesn't expect that an observer who crosses the horizon of a Schwarzschild-de Sitter or Kerr-de Sitter black hole to observe spacetime geometry becoming more and more de Sitter-like. But note that aside from this restriction, the conjectured result is a *large data result*—the conjecture is supposed

---

[47] For an important early result along these lines, see Wald, "Asymptotic Behaviour of Homogeneous Cosmological Models in the Presence of a Positive Cosmological Constant." For the state of the art (and a precise formulation of the conjecture) see Andréasson and Ringström, "Proof of the Cosmic No-Hair Conjecture in the $\mathbb{T}^3$-Gowdy Symmetric Einstein-Vlasov Setting." For a quantum cosmic no-hair result (for a quantum field theory in de Sitter spacetime), see Hollands, "Correlators, Feynman Diagrams, and Quantum No-Hair in de Sitter Spacetime." For the philosophical literature, see Doboszewski, "Interpreting Cosmic No Hair Theorems."

to hold (generically) not just for solutions that evolve from initial data that are in some sense close to de Sitter initial data, but for (geodesically complete) solutions determined by arbitrary initial data.

ii) The qualification that the conjecture holds only generically is likewise unavoidable.

The Einstein static universe is a $\Lambda > 0$ solution for a reasonable matter source that is geodesically complete, but which does not in any sense look more and more like de Sitter spacetime as $t \to \infty$: the precisely tuned matter content counterbalances the cosmological constant and there is no expansion.

Nariai spacetimes exhibit another sort of example of bad behaviour: these are geodesically complete vacuum solutions that undergo exponential expansion towards both the past and the future—without approaching de Sitter geometry in either temporal direction. Four-dimensional Nariai spacetime is the product of two-dimensional de Sitter spacetime dS_2 with the two-dimensional round sphere S_2.[48] A Nariai spacetime is homogeneous—so the only way it could be asymptotically de Sitter is if it were locally de Sitter. But since Nariai spacetimes are product spacetimes, they cannot be spaces of positive constant curvature (recall Remark 3.1).

Part of the force of the Cosmic No-Hair Conjecture is that all such troublesome spacetimes are, like the Einstein static universe, in some sense globally non-linearly unstable.[49]

QUESTION 7.7 (Larger?). The need to restrict to generic solutions is familiar from other ambitious conjectures about good behaviour in general relativity, such as the weak and strong cosmic censorship conjectures.[50] Roughly speaking, knowing that there are counterexamples to the unhedged version of the conjecture, a desirable fall-back position would be to be able to say that the conjecture holds for a set of solutions of measure one relative to some distinguished probability measure on the space of solutions. Unfortunately,

---

[48] See, e.g., Griffiths and Podolský, *Exact Space-Times*, §7.2. We can also characterize Nariai spacetime as follows: let $\mathbb{M}_6 = \{t, u, v, x, y, z\}$ be six-dimensional Minkowski spacetime, and consider the set of points (endowed with the Lorentz metric induced by the ambient Minkowski metric) that satisfy the following two conditions: $-t^2 + u^2 + v^2 = 1$ and $x^2 + y^2 + z^2 = 1$.

[49] On the instability of Nariai spacetime, see Beyer, "Non-Genericity of the Nariai Solutions. I and II."

[50] For discussion, see, e.g., Christodoulou, "On the Global Initial Value Problem and the Issue of Singularities"; Earman, *Bangs, Crunches, Whimpers, and Shrieks*, chapter 3; and Wald, " 'Weak' Cosmic Censorship."

no one has any idea how to specify such a probability measure for general relativity.[51] Possible surrogates for 'holds for a set of solutions of measure one' include 'holds except for a set of solutions of finite co-dimension', 'holds for an open and dense set of solutions', or 'holds except for a meagre set of solutions'. How satisfying are these as replacements for the measure-theoretic notion? Is it illuminating to look at how these notions function when considering questions of stability in celestial mechanics and in fluid mechanics?[52]

8. Anti-de Sitter Spacetime and Global Non-Linear Stability

In the wake of the discovery of the AdS/CFT correspondence, it was natural that attention should be turned to the question of the global non-linear stability of anti-de Sitter spacetime (AdS). This problem is interestingly different in kind from the analogous problems for Minkowski spacetime ($\mathbb{M}$) and for de Sitter spacetime (dS).

Recall from the discussion of Section 5.5 above that the conformal boundary of AdS is a timelike manifold—from which it follows that weakly asymptotically anti-de Sitter spacetimes cannot be globally hyperbolic. In this context, the natural analog of the Cauchy initial value problem for globally hyperbolic spacetimes is an initial-boundary value problem, in which standard initial data are posed on well-behaved spatial slices (that extend nicely to the conformal infinity $\mathcal{I}$) and substantive boundary conditions are imposed on $\mathcal{I}$ itself. As far as the gravitational degrees of freedom go, one natural approach is to restrict attention to strongly asymptotically anti-de Sitter spacetimes (i.e., spacetimes that admit conformal completions whose boundaries have the same topology and conformal geometry as that of AdS). This, it turns out, involves imposing *reflective* boundary conditions that rule out any flux of 'conserved' quantities across $\mathcal{I}$. In the AdS/CFT literature it

---

[51] Formidable difficulties arise already in the relatively tame setting of spatially isotropic cosmology (in which the relevant space of solutions is finite-dimensional). For discussion, see Schiffrin and Wald, "Measure and Probability in Cosmology"; and Carroll, "In What Sense Is the Early Universe Fine-Tuned?"

[52] For some discussion of the analogy with results in fluid mechanics, see Christodoulou, "Global Initial Value Problem." For an accessible introduction to some relevant results in celestial mechanics, see Dumas, *The KAM Story*.

has been customary to also impose reflective boundary conditions on any matter fields (again ruling out flux across $\mathcal{I}$).[53]

So let us first consider what is known about the question of the global non-linear stability of AdS in the context of such reflective boundary conditions.

> The proof of the non-linear stability of $\mathbb{M}$ and dS is based on a stability mechanism related to the fact that linear fields on those spacetimes satisfy sufficiently strong decay rates. The decay rates are, however, borderline in the case $\Lambda = 0$, and thus the stability of $\mathbb{M}$ is a deep fact depending on the precise non-linear structure of the [Einstein equations], whereas, in the case $\Lambda > 0$, the decay is exponential and stability can be inferred relatively easily. In contrast, on AdS it can be shown that linear fields satisfying a reflecting boundary condition on $\mathcal{I}$ remain bounded, but do *not* decay in time. It is precisely the lack of a sufficently fast decay rate at the linear level which is associated to the possibility of non-linear instability.[54]

An early result of Anderson tells us that (assuming that a certain technical hypothesis holds) if (M, g) is geodesically complete, strongly asymptotically anti-de Sitter, and both past and future asymptotically stationary, then (M, g) must in fact be anti-de Sitter spacetime.[55] So, in particular (if we impose the relevant technical assumption), no time-reflection invariant perturbation of trivial anti-de Sitter initial data determines a solution that is geodesically complete and asymptotically stationary towards the past or future.[56] Around the time this result appeared, Dafermos and Holzegel conjectured that "Anti-de Sitter space is dynamically unstable and perturbations of it lead to black holes."[57] Numerical simulations subsequently provided evidence that arbitrarily small perturbations of trivial AdS initial data should lead to the formation of black holes.[58] It has recently been

---

[53] For important exceptions, see Almheiri *et al.*, "The Entropy Bulk of Quantum Fields and the Entanglement Wedge of an Evaporating Black Hole"; and Penington, "Entanglement Wedge Reconstruction."

[54] Moschidis, *Two Instability Results in General Relativity*, 23 (I have made some changes in notation).

[55] Anderson, "On the Uniqueness and Global Dynamics of AdS Spacetimes," Corollary 1.2.

[56] Recall that initial data for the vacuum Einstein equations consist, intuitively, of the Riemannian metric and second fundamental form on a good spacelike slice. Initial data are time-reflection invariant just in case the second fundamental form vanishes. The trivial AdS initial data consist of the metric for hyperbolic space and a vanishing second fundamental form.

[57] Dafermos and Holzegel, "Dynamic Instability of Solitons in $4 + 1$-Dimensional Gravity with Negative Cosmological Constant," 2.

[58] Bizoń and Rostworowski, "Weakly Turbulent Instability of Anti-de Sitter Spacetime."

established that for the massless Einstein-Vlasov equations, arbitrarily small spherically symmetric perturbations of trivial anti-de Sitter data can determine solutions containing black holes—the physical mechanism underlying the complex proof being that the reflective boundary conditions make it possible to concentrate the energy contained in a number of distinct beams of massless particles.[59] It seems to be widely expected that this result can be extended to the asymmetric case and to other matter models.

It is far from obvious, however, that one should expect this instability to persist if the reflective boundary conditions are replaced by dissipative boundary conditions that allow flux of energy through $\mathcal{I}$.[60] Indeed, it is known that linear fields on a fixed anti-de Sitter background exhibit decay behaviour as $t \to \infty$ when subject to (maximally) dissipative boundary conditions (in marked contrast to their behaviour under reflective boundary conditions)—and it is conjectured that under suitable dissipative boundary conditions, anti-de Sitter spacetime should be globally non-linearly stable as a solution of the vacuum $\Lambda < 0$ Einstein equations.[61]

QUESTION 7.8 (Smaller?). How can we square the intense interest among physicists in AdS/CFT with the fact that for some time it has been widely suspected that anti-de Sitter spacetime is globally non-linearly unstable? Do the passages quoted in Question 7.2 need to be revisited? Is it important that no-one believes that $\Lambda < 0$ at our world?

QUESTION 7.9 (Smaller?). In AdS/CFT, the vacuum state of the boundary theory corresponds in some sense to a bulk region with an anti-de Sitter metric. The instability of AdS with reflective boundary conditions tells us that even boundary states arbitrarily close to the vacuum state evolve to boundary states that correspond to black holes in the bulk. How do we reconcile this picture with the fact that since dynamical evolution in the boundary theory is unitary, a state initially close to the vacuum state (which is itself invariant under time evolution) can only evolve to states just as close to the vacuum state?

---

[59] Moschidis, "A Proof of the Instability of AdS for the Einstein-Massless Vlasov System." The solutions in question have complete conformal boundaries—see Moschidis, "The Characteristic Initial-Boundary Value Problem for the Einstein-Massless Vlasov System in Spherical Symmetry," §2.2 and appendix B.

[60] This point is emphasized in Friedrich, "On the AdS Stability Problem."

[61] See Holzegel *et al.*, "Asymptotic Properties of Linear Field Equations in Anti-de Sitter Space."

8

Cosmic Topology

1. Introduction

Glymour introduced philosophers to the possibility that difficulties of principle might stand in the way of attempts to determine the topology of space or spacetime on the basis of observational evidence:

> for each of a class of fashionable cosmological models there is another (unfashionable) model different from the first in the topology it ascribes to space-time, and there are good reasons to think that any two such cosmological models are, both in fact and in principle, experimentally indistinguishable. Any bit of evidence which we can account for with one model, we can account for with another, and conversely.[1]

In the subsequent literature, the phenomenon of 'indistinguishable space-times' has often been taken to place an interesting upper bound on scientific knowledge—and hence to have crucial relevance to debates over scientific realism.[2]

One of the themes of Glymour's papers is that expanding cosmological models provide interesting examples in which the topology of space cannot be determined by observation, even in the limit of infinitely large data sets. This theme was noted in a paper of Malament that further develops Glymour's approach:

> In some space-time models studied in relativity theory any particular observer can receive signals from, and hence directly acquire information

---

[1] Glymour, "Topology, Cosmology, and Convention," 195 f. Some related points were raised in the physics literature around the same time—for references, see Glymour, "Indistinguishable Space-Times and the Fundamental Group," 50.

[2] See, e.g., Glymour, *Theory and Evidence*, 354 ff.; van Fraassen, *An Introduction to the Philosophy of Time and Space*, postscript §1; Earman, "Underdetermination, Realism, and Reason," §9; Stanford, *Exceeding Our Grasp*, 13 f.; and Manchak, "Can We Know the Global Structure of Spacetime?" For a dissenting voice, see Norton, "Observationally Indistinguishable Spacetimes."

Accelerating Expansion: Philosophy and Physics with a Positive Cosmological Constant. Gordon Belot, Oxford University Press. © Gordon Belot 2023. DOI: 10.1093/oso/9780192866462.003.0009

about, only a limited region of space-time. This happens, for instance, in a rapidly expanding universe in which galaxies that might try to signal one another are actually receding from one another at velocities approaching that of light. It may turn out in these cases that the information from that limited region of space-time which any one observer can have access to is compatible with quite different overall space-time structures.[3]

However, in the literature inspired by Glymour's and Malament's papers, there has been a diminishing focus on the connection between rapid expansion and indistinguishability (and, more generally, on examples of the sort that arise in cosmology). Here I seek to institute a back to Glymour movement by redirecting attention towards varieties of indistinguishability endemic to cosmology.

Section 2 below reviews the central philosophical notions of indistinguishability due to Glymour and to Malament. Section 3 offers a quick survey of some basic facts about the workhorse models of cosmology. Section 4 considers how Glymour's basic definition functions when applied to spatially flat cosmological models (a case of central interest according to our current view of the Universe). The decision to work with such models involves an implicit topological assumption—in Section 5 we discuss just how strong this assumption is. The moral of the story told in these initial sections will be that within the standard framework of cosmological modelling, we expect that no matter how much data we collect, a small handful of spatial topologies will always be consistent with observation. So observation and our theoretical presuppositions cannot determine the topology of space—but they *almost* do so. Section 6 is devoted to exposition of a result due to Ringström that overturns this picture. It can be thought of as saying that if we replace Glymour's notion of observational indistinguishability by a natural merely approximate analog, then there is a sense in which observation can place *no* constraints on possible (compact) spatial topologies.

2. Observational Indistinguishability

Recall that if (M, g) is a time-oriented Lorentz manifold and S is a set of points of M, then $I^-(S)$, the *chronological past of S*, is the set of points

---

[3] Malament, "Observationally Indistinguishable Space-Times," 61.

$p \in M$ such that there exists a future-directed timelike curve from p to a point in S.

DEFINITION 8.1 (Glymour). Time-oriented Lorentz manifolds (M, g) and (M', g') are *observationally indistinguishable* if for every inextendible timelike curve γ_1 in one, there is an inextendible timelike curve γ_2 in the other, such that $I^-(\gamma_1)$ and $I^-(\gamma_2)$ are isometric.

Some Lorentz manifolds are observationally indistinguishable only from (manifolds isometric to) themselves. Minkowski spacetime is an example: if γ is an inextendible timelike curve in Minkowski spacetime, then $I^-(\gamma)$ is all of Minkowski spacetime; so any Lorentz manifold (M', g') observationally indistinguishable from Minkowski spacetime must contain a region R isometric to Minkowski spacetime—from which it follows that R must be all of M' (see Exercise 8.1 below).

But it is possible for spacetimes with quite different geometries to be observationally indistinguishable from one another.

EXAMPLE 8.1. Consider a two-dimensional de Sitter spacetime dS_2.[4] Recall from Remark 3.2 above that for each $k = 2, 3, \ldots$ the k-fold cover, $dS_2^{(k)}$, of dS_2 is the result of taking k copies of dS_2, cutting each open along a timelike geodesic, and then sewing them together to form a spacetime with the same topology and local geometry as dS_2, but in which the Cauchy surfaces of minimum volume are k times larger than in dS_2. Recall also that by sewing together along their edges a countable infinity of copies of dS_2, each of which has been cut open along a timelike geodesic, we can construct the universal cover $\widetilde{dS_2}$ of dS_2, which has the local geometry of dS_2, the topology of $\mathbb{R}^2$, and infinite-volume Cauchy surfaces. Each of $dS_2, dS_2^{(2)}, \ldots, \widetilde{dS_2}$ is observationally indistinguishable from each other.

EXAMPLE 8.2. Let C be the strip $\{(t, x) \mid t \in \mathbb{R}, x \in [-1, 1]\}$.[5] Consider two Lorentz metrics we might put on C:

$$g_0 = -dt^2 + (\cosh^2 \pi t) dx^2$$
$$g_1 = -dt^2 + (2 - x^2)(\cosh^2 \pi t) dx^2$$

---

[4] For this example, see Malament "Observationally Indistinguishable Space-Times," 64 f. For further discussion, see Norton, "Observationally Indistinguishable Spacetimes."
[5] For this example, see Malament "Observationally Indistinguishable Space-Times," 66 f.

Here g_0 is the two-dimensional de Sitter metric and g_1 is the result of deforming the spatial part of the geometry of g_0 by a fudge factor. Construct a Lorentz manifold M by taking a copy of C equipped with g_0 and another equipped with g_1, and identifying their edges in the obvious way so that the resulting manifold has topology $\mathbb{R} \times S^1$. Let $\widetilde{M}$ be the universal cover of M: the manifold $\mathbb{R} \times \mathbb{R}$ equipped with a metric consisting of alternating copies of g_0 and g_1. M and $\widetilde{M}$ are observationally indistinguishable.[6] Each is geodesically complete, globally hyperbolic, and a solution of the vacuum Einstein equations.[7]

EXERCISE 8.1 (Medium). Let (M, g) be Lorentz manifold. Recall that we say that (M, g) *extendible* if it is isometric to a proper open subset of some other Lorentz manifold $(N, h.)$ And we say that (M, g) is *complete* if each of its inextendible geodesics admits an affine parameterization that ranges over the complete real line. Show that every complete Lorentz manifold is inextendible.[8]

EXERCISE 8.2 (Medium). Explain why the restriction to two spacetime dimensions was essential in Example 8.1.[9]

EXERCISE 8.3 (Easier). Think about how to re-tool the construction of Example 8.1 to apply to four-dimensional Nariai spacetimes.[10]

EXERCISE 8.4 (Easier). Use a variant on the construction of Example 8.2 above to construct an uncountable family of pairwise observationally indistinguishable but non-isometric Lorentz manifolds.[11]

<hr/>

[6] Any observer in either sees just the geometry of g_0, just the geometry of g_1, or a geometry that includes a seam between a g_0-region and a g_1-region. See Malament, "Observationally Indistinguishable Space-Times," 65 f.

[7] The last point is not impressive: in two spacetime dimensions, *every* Lorentz metric is a solution of the vacuum Einstein equations. For discussion of this and other peculiarities of general relativity in two dimensions, see Fletcher *et al.*, "Would Two Dimensions Be World Enough for Spacetime?"

[8] HINT: suppose otherwise and consider a boundary point of a region isometric to (M, g) in (N, h).

[9] HINT: topology.

[10] Recall that Nariai spacetime is the direct product of a round two-sphere with a two-dimensional de Sitter geometry—see Griffiths and Podolský, *Exact Space-Times in Einstein's General Relativity*, §7.2.

[11] HINT: there are uncountably many non-constant maps from the integers to $\{0, 1\}$.

EXERCISE 8.5 (Medium). Think about how to exploit features of the Schwarzschild-de Sitter family of solutions to construct a countably infinite family of pairwise observationally indistinguishable and non-isometric Lorentz manifolds.[12]

EXERCISE 8.6 (Easier). Malament shows that various global features of Lorentz manifolds are preserved under observational equivalence.[13]

a) Let X be a property of Lorentz manifolds and suppose that X can be given a characterization of the form: (M, g) fails to have X if and only if there is a point p in M such that $I^-(p)$ has geometric feature Φ (i.e., whether or not $I^-(p)$ has Φ depends only on the geometry that g induces on $I^-(p)$). Show that if (M, g) lacks X, then so does any Lorentz manifold observationally indistinguishable from (M, g).

b) A Lorentz manifold is *causal* if it contains no closed causal curves. Show that if (M, g) is not causal, then neither is any Lorentz manifold observationally indistinguishable from (M, g).

c) A Lorentz manifold is *diamond-compact* if for any points p and q in the manifold, the set $J^+(p) \bigcap J^-(q)$ is compact.[14] Show that if (M, g) is not diamond-compact, then neither is any Lorentz manifold observationally indistinguishable from (M, g).

d) Let (M, g) and (M', g') be observationally indistinguishable Lorentz manifolds. Show that both are causal or neither is. Show that both are diamond-compact or neither is.

Note that a Lorentz manifold fails to be globally hyperbolic if and only if it fails to be causal or fails to be diamond-compact.[15] So the above establishes

---

[12] HINT: for relevant background, see Griffiths and Podolský, *Exact Space-Times*, §9.4. For a discussion of topology, see Beig and Heinzle, "CMC-Slicings of Kottler-Schwarzschild-de Sitter Cosmologies." For a way of including matter content, see Andréasson, Fajman, and Thaller, "Static Solutions to the Einstein-Vlasov System with a Non-vanishing Cosmological Constant."

[13] Malament, "Observationally Indistinguishable Space-Times," §II.

[14] Here $J^+(p)$ is the set of points that can be reached by future-directed causal curves departing from p and $J^-(q)$ is the set of points that can be reached by past-directed causal curves departing from q.

[15] See Bernal and Sánchez, "Globally Hyperbolic Spacetimes Can Be Defined as 'Causal' Instead of 'Strongly Causal.'" In fact, if we set aside the weirdo cases of two-dimensional Lorentz manifolds and compact Lorentz manifolds, then global hyperbolicity is equivalent to diamond-compactness—see Hounnonkpe and Minguzzi, "Globally Hyperbolic Spacetimes Can Be Defined without the 'Causal' Condition."

that for any two observationally indistinguishable Lorentz manifolds, either both are globally hyperbolic or neither is.

EXERCISE 8.7 (Harder). We call a Lorentz manifold (M, g) *standard static* if there is a Riemannian manifold (N, h) and a smooth $f : N \to \mathbb{R}^+$ such that (M, g) is isometric to $(\mathbb{R} \times N, -f \, dt^2 + h)$.[16] Show that any standard static (M, g) is observationally indistinguishable only from itself.[17]

2.1 A Worry about Observational Indistinguishability

In a discussion of observational indistinguishability, Manchak remarks that

> Intuitively, one space-time is observationally indistinguishable from another if no observer in the first space-time has grounds for deciding which of the two she inhabits.[18]

He continues:

> It should be clear from the definition that, given that a space-time (M, g_{ab}) is observationally indistinguishable from another space-time (M', g'_{ab}), no empirical data could (even in principle) allow an observer in (M, g_{ab}) to distinguish between the two models.

These remarks deftly encapsulate the way that Malament and Glymour, too, seem to think of the epistemological significance of observational indistinguishability (and its variants).

It is important to note that strong epistemological assumptions are in play here. Consider the following variant on Example 8.2 above. Recall that there

---

[16] For a discussion of standard static spacetimes, see Sánchez, "On the Geometry of Static Spacetimes." Note that every standard static Lorentz manifold is complete and static but that there exist Lorentz manifolds that are complete, static, and globally hyperbolic without being standard static.

[17] HINT: see Straumann, *General Relativity*, §2.7 for an analysis of the geometry of light rays in (M, g) in terms of the geodesics of the Riemannian manifold $(N, \frac{1}{f(x)} g(x))$. Let ℓ be one of the orbits of the group of time translation symmetries of (M, g) and show that $I^-(\ell) = M$.

[18] Manchak, "What Is a Physically Reasonable Space-Time?", 412; see also §5 of his *Global Spacetime Structure*. Warning: in the passage quoted, Manchak has in mind not Glymour's notion of observational indistinguishability, but a weaker notion due to Malament, *weak observational indistinguishability* (see Section 2.2 below). If the worry developed here applies to Glymour's notion it also applies to all weaker forms of observational indistinguishability.

C was a vertical strip in $\mathbb{R}^2$; g_0 was a de Sitter metric on C; and g_1 was a metric that differed from g_0 in the central region of the strip but agreed with it along the edges. We can endow $\mathbb{R}^2$ with a Lorentz metric by suitably pasting together copies of (C, g_0) and (C, g_1). Let us call N_0 the result of employing a countable infinity of copies of (C, g_0) and a single copy of (C, g_1). And let us call N_1 the result of employing a countable infinity of copies of (C, g_1) and a single copy of (C, g_0). On the Glymour-Malament-Manchak view, observing a two-dimensional de Sitter geometry provides no reason to think that you live in a world with geometry N_0 rather than geometry N_1—after all, both worlds contain regions with de Sitter geometry. But there is a competing way of thinking about this sort of case. Intuitively, in worlds with geometry N_0, typical observers see de Sitter geometry and only exceptional observers see something else—and the situation is reversed in worlds with geometry N_1. A guiding thought of cosmological theorizing is that we should think that we are typical rather than exceptional.[19] On this way of thinking, seeing de Sitter geometry would provide very strong reason to think that we live in N_0 rather than N_1.

It is not just cosmologists who like to think this way. This fact is sometimes obscured in the philosophical literature on self-locating belief by a proliferation of rococo examples (involving mad scientists, memory erasing drugs, etc.). But the key point can be made without appeal to such exotica.

For instance, suppose that you and 20 friends have booked all the rooms in a 21-room hotel. You remember either reading in the hotel brochure that all but one room in the hotel is red, or reading that only one room is red. You just do not remember which, and you are initially about 50-50 as to the type of hotel that you have booked. Upon arrival you and your friends randomly pick rooms to go to. You then find that your room is red. If you took your evidence to be the qualitative proposition *Someone is in a red room* you would have no reason to modify your credences regarding the type of hotel that you are in, since that proposition had to be true either way. But if your evidence is the self-locating proposition *I am in a red room*, it strongly supports the hypothesis that all but one room in your hotel is red. And it seems obvious that this is the correct way to reason. This is

---

[19] Example: at the scales relevant to cosmology, the night sky looks the same in every direction from Earth; this motivates cosmologists to look for models of the Universe in which the night sky looks roughly the same in every direction *at each point*, rather than to look for models in which there is at least one point with this feature. See Section 3.1 below for further discussion.

presumably what you would conclude if you believed that you were the only person in the hotel, and it surely makes no relevant difference whether you believe that you have 20 friends with you or not.[20]

QUESTION 8.1 (Smaller). Determine which, if any, of the examples and results discussed in this chapter are of interest if you take seriously the above worry about typicality.

QUESTION 8.2 (Larger?). What sort of epistemology is implicit in the way that Manchak, Malament, and Glymour are thinking of observational indistinguishability? Is it sustainable? If not, what should it be replaced with? What technical notions of indistinguishability are most interesting, epistemologically?

QUESTION 8.3 (Larger). Make the best case that you can against the line pushed in the Dorr and Arntzenius quote above. (Suggestion: you may find it helpful to consult the literature cited in fn. 77 of Chapter 9 below.)

2.2 Variants on Observational Indistinguishability

Before getting down to applications of Glymour's notion to cosmological examples, let us look quickly at a couple of variants on it that have been widely discussed among philosophers.

Our first variant definition is implicit in Glymour's discussion.[21] Malament codifies it and motivates it as follows:

> As Glymour's definition is formulated, space-times can be observationally distinguishable from each other without an observer in either one necessarily being able to distinguish between them *at any time* during the course of his life. It is sufficient that the composite, lifelong, integrated knowledge of one observer distinguish between them. A weaker condition of observational indistinguishability which Glymour considers insists that observational distinction between space-times be made *within* the lifetime of some observer.[22]

---

[20] Dorr and Arntzenius, "Self-Locating Priors and Cosmological Measures," 399.
[21] See Glymour, "Indistinguishable Space-Times," 53.
[22] Malament, "Observationally Indistinguishable Space-Times," 66.

This train of thought leads us to the following:

DEFINITION 8.2. Time-oriented Lorentz manifolds (M, g) and (M', g') are *observationally indistinguishable at finite times* if for every point in one of them, there is a point in the other, such that the chronological pasts of the two points are isometric.

Observational indistinguishability clearly implies observational indistinguishability at finite times. But the converse does not hold.

EXAMPLE 8.3 (Malament). Take a copy of Minkowski spacetime, choose inertial coordinates, and delete all points at all times $t \geq 0$. The resulting spacetime is observationally indistinguishable at finite times from Minkowski spacetime: in either, the chronological past of any point is isometric to the chronological past of the origin in Minkowski spacetime. But they are not observationally indistinguishable: in Minkowski spacetime, the chronological past of any inextendible observer is all of Minkowski spacetime; no observer in truncated Minkowski spacetime can have a chronological past of this form.

Malament proposed a yet weaker notion of indistinguishability.

DEFINITION 8.3. Let (M, g) and (M', g') be Lorentz manifolds. Then (M', g') is *weakly observationally indistinguishable* from (M, g) if for every point in (M, g) there is a point in (M', g'), such that the chronological pasts of the two points are isometric.

Consider two-dimensional de Sitter spacetime and the spacetime M from Example 8.2 above (built by gluing together two vertical strips from the plane, one of which has de Sitter geometry). There are points in M whose chronological pasts do not look like pieces of de Sitter spacetime. So M is not weakly observationally distinguishable from de Sitter spacetime. But de Sitter spacetime is weakly observationally indistinguishable from M: the chronological past of any point in de Sitter spacetime looks just like the chronological past of some point in M (the points along the axis of symmetry of the de Sitter strip used to construct M can see only the de Sitter portion of the geometry). So weak observational indistinguishability is not a symmetric relation—and it is strictly weaker than observational indistinguishability at finite times.

We say that a Lorentz manifold (M, g) is *causally bizarre* if there is a point in M whose chronological past is all of M. And we say that a Lorentz manifold is *weakly observationally indistinguishable only from itself* if all Lorentz manifolds weakly observationally indistinguishable from (M, g) are isometric to (M, g).

Malament conjectured that in order for a Lorentz manifold to be weakly observationally indistinguishable only from itself, it must be causally bizarre. And he sketched a recipe that when given as input a non-causally bizarre Lorentz manifold (M, g) gives as output a Lorentz manifold (M', g') that is not isometric to (M, g) but from which (M, g) is weakly observationally indistinguishable.[23] Manchak subsequently gave a detailed proof of this result (via a distinct recipe of this type).[24]

It is surprising and fascinating that weak observational indistinguishability is so common. But it is important to note that known examples of Lorentz manifolds that are weakly observationally indistinguishable from, e.g., Minkowksi spacetime would all count as irrelevant monstrosities by the standards of working cosmologists.[25] Applying Malament's construction to Minkowski spacetime yields a flat Lorentz manifold in which every causal geodesic is future-incomplete (so every observer eventually just pops out of existence). Manchak's construction yields a flat manifold in which every observer initially sees an infinite past that looks just like Minkowski spacetime, but eventually sees naked singularities or is destroyed by one—further, this manifold has infinitely many flat asymptotic ends (disjoint regions, each of which looks like Minkowski spacetime to the exterior of a compact set).

So the dialectical relevance to the question of scientific realism of results concerning weak observational indistinguishability is highly contestable. Consider the truncated version of Minkowski spacetime used in Example 8.3 above. This is liable to strike working cosmologists as being about as relevant to their business as geologists and palaeontologists consider young Earth creationism—the hypothesis that the world was created 5,000 years ago, replete with various traces of a much longer history—to theirs. Maybe they are wrong to be so dismissive—that is an interesting question. But part of the point of the present discussion is to convince readers interested in the bearing of indistinguishable spacetimes on scientific realism that it is a question that they do not need to enter into. We can find interesting examples

---

[23] See Malament, "Observationally Indistinguishable Space-Times," 69.
[24] Manchak, "Can We Know?," §4.
[25] On this point, see Cinti and Fano, "Careful with Those Scissors, Eugene!"

of underdetermination of topology by all possible evidence without straying beyond possibilities taken seriously in contemporary cosmology.

EXERCISE 8.8. This is a continuation of Exercise 8.6 above.[26]

a) (Easier). Suppose that (M,g) is weakly observationally indistinguishable from (M',g'). Show that: if (M,g) is not causal, then neither is (M',g'); that if (M,g) is not diamond-compact, then neither is (M',g'); that if (M,g) is not globally hyperbolic, then neither is (M',g').

b) (Easier). Show that if two Lorentz manifolds are observationally indistinguishable at finite times, then either both are causal or neither is, either both are diamond-compact or neither is, and either both are globally hyperbolic or neither is.

c) (Medium). Show that from the fact (M,g) is weakly observationally indistinguishable from (M',g') it does not follow that: (M',g') must be causal if (M,g) is; (M',g') must be diamond-compact if (M,g) is; (M',g') must be globally hyperbolic if (M,g) is.

EXERCISE 8.9 (Harder). Show that among complete Lorentz geometries, Minkowski spacetime is observationally indistinguishable at finite times only from itself—i.e., if (M,g) is complete and observationally indistinguishable at finite times from Minkowski spacetime, then it is isometric to Minkowski spacetime.

EXERCISE 8.10 (Easier). Given an example of an incomplete Lorentz manifold that is weakly observationally indistinguishable from the Einstein static universe. Given an example of a complete but non-static Lorentz manifold with this feature.[27]

EXERCISE 8.11 (Harder). Let (M,g) be a standard static Lorentz geometry.[28] Show that among static Lorentz geometries, (M,g) is weakly observationally indistinguishable only from itself.[29]

---

[26] We are again following in the footsteps of Malament, "Observationally Indistinguishable Space-Times," §II.

[27] HINT: for the first part, adapt Example 8.3 above; for the second, adapt the example that Malament uses to show that global hyperbolicity need not be preserved under weak observational equivalence—see his "Observationally Indistinguishable Space-Times," 74.

[28] For this notion, see Exercise 8.7 above.

[29] HINT: recall that as we have set things up, each static Lorentz geometry admits a complete timelike Killing field K (alternative conventions allow K to be incomplete or to be timelike only in a neighbourhood of spacelike and null infinity). Fix K. Choose a leaf Σ_0 of the foliation by

3. Cosmological Models

Here we review some basic facts about standard cosmological models—often called *Friedmann-Robertson-Walker models* or *Friedmann-Lemaître-Robertson-Walker models*.[30] For convenience, we will make a separation between kinematics-geometry (which we associate with Robertson and Walker) and dynamics (which we associate with Friedmann and Lemaître).[31]

3.1 Robertson-Walker Manifolds

From where we sit here and now, the Universe looks pretty well the same in every direction on cosmological scales: the statistical distribution of galaxies is much the same in each direction; the temperature of the cosmic background radiation even more so. Bondi suggested the name *the Copernican Principle* for the thesis "that the Earth is not in a central, specially favoured position"—and then remarked that "it is only a small step from this principle to the statement that the Earth is in a *typical position*."[32] Something like the latter thought is usually taken as basic in modern cosmology. So the fact that from here the Universe looks much the same in every direction is taken to warrant the so-called *Cosmological Principle*: that at any instant of time, the Universe is homogeneous and isotropic (at cosmological scales).[33] So it is natural to be interested in cosmological models that admit a decomposition into instants of time at each of which geometry and matter appear to be isotropic (and hence also homogeneous) relative to the distinguished observers whose worldlines are orthogonal to these instants of time.[34]

As we saw in Section 3.3 above, the only isotropic three-dimensional Riemannian manifolds are the hyperbolic, Euclidean, spherical, and elliptic

spacelike surfaces orthogonal to K. Let p_0 be a point on Σ_0 and show that the geometry of Σ_0 can be determined by looking at the chronological pasts of points along the integral curve of K that passes through p_0. Use this to constrain the possible static Lorentz geometries weakly observationally indistinguishable from (M, g). For background on static spacetimes, consult Sánchez, "Geometry of Static Spacetimes."

[30] For an overview of the evidence for empirical claims made in this section, see, e.g., Dodelson and Schmidt, *Modern Cosmology*, chapter 1; or Peebles, *Cosmology's Century*.

[31] Our nomenclature loosely follows that of Choquet-Bruhat, *Introduction to General Relativity, Black Holes, and Cosmology*, chapter VII.

[32] Bondi, *Cosmology*, 13.

[33] On the history of (and relations between) the Copernican Principle and the Cosmological Principle, see Beisbart and Jung, "Privileged, Typical, or Not Even That?"

[34] Recall that isotropy implies homogeneity for Riemannian manifolds—see, e.g., Wolf, *Spaces of Constant Curvature*, lemma 11.6.6.

spaces. So requiring our cosmological models to be spatially isotropic would place very strong constraints on the topology of space: it would have to have the topology of $\mathbb{R}^3$ (hyperbolic or Euclidean), S^3 (spherical), or $S^3/\mathbb{Z}_2$ (elliptic). But such a restriction would not follow from the line of reasoning sketched above. Observation tells us only that the portion of the Universe observable from here looks much the same in every spatial direction. Adding the principle that we are in no way special gets us that from every location, the observable portion of the Universe should look much the same in every spatial direction. So rather than requiring our models to fall into instants of time modelled by isotropic Riemannian manifolds, we should require only that they fall into instants of time modelled by *locally* isotropic Riemannian manifolds.[35] And for three-dimensional Riemannian manifolds, being locally isotropic is equivalent to having constant curvature. So this opens up many possible spatial topologies beyond tame ones just mentioned.[36]

In cosmology it is natural to work with globally hyperbolic models. That makes it natural to require spatial sections to be geodesically complete—if we allowed incomplete spatial sections, then observers could be snuffed out by entering the wrong part of space, so there would be inextendible causal curves that did not register at each instant of time.[37]

DEFINITION 8.4. A *space form* is a complete Riemannian n-manifold ($n \geq 2$) of constant curvature $k = -1, 0,$ or 1.

DEFINITION 8.5. Let I be a (possibly infinite) open interval of real numbers, let (Σ, h) be a three-dimensional space form, and let $a : I \to (0, \infty)$ be a smooth function (called the *scale factor*). Let $M = I \times \Sigma$ and let $t : M \to I$ be the projection map $t : (\tau, p) \mapsto \tau$. Set $g := -dt^2 + a^2(t)h$. We call the Lorentz manifold (M, g) a *Robertson-Walker manifold*.

---

[35] Of course, it would also be natural to be interested in models that were merely approximately locally isotropic in some appropriate sense. Hold that thought until Section 6 below.

[36] The possible cosmological relevance of spaces of constant curvature with less tame topologies was advocated by Klein already in the nineteenth century—on Klein and his influence on Killing and Schwarzschild, amongst others, see Epple, "From Quaternions to Cosmology." In the cosmological literature, this possible relevance is gestured at already in Friedmann, "On the Possibility of a World with Constant Negative Curvature of Space," §3.1. While this possibility is sometimes glossed over in textbook treatments of cosmology, it is typically at least mentioned in general relativity textbooks (even if only to be dismissed as unmotivated)—see, e.g., Misner, Thorne, and Wheeler, *Gravitation*, box 27.2; Hawking and Ellis, *The Large Scale Structure of Space-Time*, 136; Wald, *General Relativity*, 95; and Carroll, *Spacetime and Geometry*, 331.

[37] See Beem, Ehrlich, and Easley, *Global Lorentzian Geometry*, theorem 3.66.

Let $(I \times \Sigma, -dt^2 + a^2(t)h)$ be a Robertson-Walker manifold with h a metric of constant curvature k. Fixing a value $t \in I$ determines a Cauchy surface

$$\Sigma_t := \{(t,x) \,|\, x \in \Sigma\}$$

of topology Σ. The Riemannian metric that g induces on Σ_t has constant sectional curvature $k/a^2(t)$. Fixing a point $x \in \Sigma$ determines a timelike geodesic of the form $\gamma_x : t \in I \mapsto (t,x)$. Each such curve is orthogonal to each of the level surfaces Σ_t. We call these distinguished geodesics *fundamental observers*. By exchanging light signals and applying a local version of the Einstein simultaneity convention, the fundamental observers can single out the hypersurfaces Σ_t (the level surfaces of t) as empirically distinguished instants of time.[38] Each fundamental observer sees the sky and the geometry of space as being (locally) the same in each direction at each moment of time.

In a Robertson-Walker geometry $(I \times \Sigma, -dt^2 + a^2(t)h)$, specifying a fundamental observer is the same thing as specifying a point of Σ. So we can think of Σ as the space of fundamental observers, equipped with a natural Riemannian metric h. A ball $B_r(\gamma) \subset \Sigma$ of radius r centred at γ in Σ consists of a cloud of fundamental observers. Each observer in the cloud determines a timelike geodesic in $I \times \Sigma$ and collectively they determine a world tube (centred on γ), the intersection of which with a distinguished instant of time Σ_t will be a sphere of radius $r \cdot a(t)$.

EXERCISE 8.12. The fastidious will worry that in order to pick out the fundamental observers and the distinguished Cauchy surfaces Σ_t of a Robertson-Walker manifold (M, g), we use more than the structure of M as a manifold and g as a Lorentz metric—we need to pick a distinguished way of viewing M as a product manifold (with g adapted to this product structure).

  a) (Easier). Show that the fastidious are right: give an example of a Robertson-Walker manifold (M, g) and an isometry $d : M \to M$ such that the family of fundamental observers of (M, g) is not invariant under d.[39]

  b) (Medium). Restrict attention to simply connected Robertson-Walker manifolds—ones where (Σ, h) is hyperbolic, Euclidean, or

---

[38] For this construction, see, e.g., Sachs and Wu, *General Relativity for Mathematicians*, §5.3.
[39] HINT: use the most highly symmetric Robertson-Walker manifold you can think of.

spherical three-space. Construct examples of such Robertson-Walker manifolds that have six-dimensional isometry groups (the minimum size possible). Construct others that have ten-dimensional isometry groups (the maximum size possible). Find an example with isometry group of intermediate size—and try to compile a complete list of such examples, putting constraints on allowed matter content if necessary.[40]

c) (Medium). Isolate a sense in which the Worry of the Fastidious doesn't arise for generic Robertson-Walker manifolds.

3.2 Friedmann-Lemaître Solutions

So far we are just talking about geometry—we have not imposed the Einstein equations. The most basic cosmological models use perfect fluids to model the radiation and massive matter that fill the Universe.[41] In the setting of Robertson-Walker manifolds, it is natural to require the fluids to be spatially homogeneous and isotropic according to the fundamental observers. The stress-energy for such a fluid takes the form:

$$T_{ab} = \rho u_a u_b + p g_{ab},$$

where $\rho(t)$ and $p(t)$ are the (spatially homogeneous) mass-energy density and pressure of the fluid and u is the four-velocity of the fundamental observers.[42] Different types of fluid correspond to different equations of state (i.e., relations between pressure and density). We will model massive matter as a pressure-free dust, imposing the equation of state $p_d \equiv 0$. And we will use the equation of state $p_r = \rho_r/3$ to model a radiation fluid (consisting of massless particles).

DEFINITION 8.6. A *Friedmann-Lemaître solution* is a Robertson-Walker manifold (M, g) containing a dust with density $\rho_d(t)$ and a radiation fluid with density $\rho_r(t)$, satisfying the Einstein-Euler equations (possibly with $\Lambda \neq 0$).

---

[40] HINT: we have already repeatedly mentioned some relevant examples; for helpful background, consult Patrangenaru, "Lorentz Manifolds with the Three Largest Degrees of Symmetry"; and chapters 11 and 12 of Stephani *et al.*, *Exact Solutions of Einstein's Field Equations*.

[41] For discussion, see Misner, Thorne, and Wheeler, *Gravitation*, §§27.2 and 27.7.

[42] See, e.g., Sachs and Wu, *General Relativity for Mathematicians*, §3.15.

Note that we allow the cases $\rho_d \equiv 0$ or $\rho_r \equiv 0$, so we are in effect also allowing vacuum solutions and solutions including just dust or just radiation.

In the present setting, the Einstein-Euler equations become

$$3\left(\frac{\dot{a}^2 + k}{a^2}\right) = \rho_d + \rho_r + \Lambda$$

$$2\frac{\ddot{a}}{a} + \left(\frac{\dot{a}}{a}\right)^2 + \frac{k}{a^2} = \Lambda - \frac{1}{3}\rho_r$$

$$\rho_d(t)a^3(t) \equiv \text{const.}$$

$$\rho_r(t)a^4(t) \equiv \text{const.}$$

where the first pair of equations are the Einstein equations for our stress-energy tensor and the second pair state that this stress-energy tensor is covariantly constant.[43] Note that if $a(t)$ is monotonically increasing in t, then the density of the radiation fluid dilutes much more rapidly than the density of the dust—the heuristic explanation being that if the scale factor doubles in size, the number of particles per unit volume is diminished by a factor of eight, while the energy density of photons will be diluted by a further factor of two (the number of photons per unit volume is diminished—and their wavelength is stretched).[44]

Observation pushes us towards the following picture of our Universe at cosmological scales. It is a two-fluid Friedmann-Lemaître solution of the above type with a positive cosmological constant. It began with a singularity and will expand permanently towards the future, so we can take $I = (0, \infty)$. It appears to be spatially flat ($k = 0$). The dynamics was dominated at early times by the energy density of radiation, which implies that $a(t)$ behaves roughly as $t^{1/2}$ at such times. At late times, the cosmological constant dominates and $a(t)$ behaves roughly as e^t.[45] We will call Friedmann-Lemaître solutions of this type *standard cosmological models*.

EXERCISE 8.13 (Easier). A *static* Friedmann-Lemaître solution is one for which $I = \mathbb{R}$ and the scale factor $a(t)$ is a constant function. Show that there

---

[43] See, e.g., Ringström, *On the Topology and Future Stability of the Universe*, §28.1; and Choquet-Bruhat, *Introduction to General Relativity, Black Holes, and Cosmology*, §VII.6.1. Note that in this setting we do not require separate equations of motion for the matter content of the universe—on this point, see, e.g., Thorne and Blandford, *Modern Classical Physics*, §28.2.4.

[44] On this point, and for its bearing on the intuitive notion of the conservation of energy, see Carroll, *Spacetime and Geometry*, 119 f.

[45] For precise asymptotics, see Ringström, *Topology and Future Stability*, §28.2.

are no static Friedmann-Lemaître solutions in the $k = -1$ case; that when $k = 0$, static Friedmann-Lemaître solutions are vacuum $\Lambda = 0$ solutions; and that when $k = 1$, static Friedmann-Lemaître solutions involve nontrivial matter and a positive cosmological constant.

4. Cosmology and Indistinguishability

We are ready to apply Glymour's apparatus to the sort of examples that arise in cosmology.

To this end, let us specialize to the $k = 0$ case. We consider Robertson-Walker manifolds of the form $(I \times \Sigma, -dt^2 + a^2(t)h)$, where (Σ, h) is a *flat* three-dimensional space form. We call such a Robertson-Walker manifold *Euclidean* if (Σ, h) is $\mathbb{E}^3$ ($= \mathbb{R}^3$ with the standard Pythagorean formula metric). In a Euclidean Robertson-Walker manifold, we can choose coordinates so that the line element is given by:

$$ds^2 = -dt^2 + a^2(t)\left[dx_1^2 + dx_2^2 + dx_3^2\right],$$

where $x_1, x_2, x_3 \in \mathbb{R}$ and $t \in I$. We call such coordinates *canonical*. We denote by γ_0 the fundamental observer at rest at the origin of $\mathbb{E}^3$.

Fix a Euclidean Robertson-Walker manifold and a set of canonical coordinates for it. Choose two times $t_1, t_2 \in I$. We are interested in the question: which fundamental observers can send light signals at time t_1 that will reach γ_0 no later than t_2? Given the symmetries of the problem, we know that the fundamental observers who can perform this feat will form a ball in the space of observers, centred at γ_0 and with radius $\hat{R} \in [0, \infty]$. In order to find $\hat{R}$, suppose that a signal is sent from event $p_1 = (t_1, x_1, y_1, z_1)$ to event $p_2 = (t_2, 0, 0, 0)$ on the worldline of the fundamental observer γ_0. Switch to spherical coordinates for Σ and parameterize the path taken by the signal as $\sigma(t) = (t, r(t), \theta(t), \varphi(t))$ ($t \in [t_1, t_2]$). Since the signal is causal, we know that:

$$-1 + a^2(t)|\dot{r}(t)|^2 + a^2(t)|\dot{\theta}(t)|^2 + a^2(t)|\dot{\varphi}(t)|^2 \leq 0.$$

So, in particular,

$$-1 + a^2(t)|\dot{r}(t)|^2 \leq 0.$$

It follows that:

$$\hat{R} = \int_{t_1}^{t_2} |\dot{r}(t)|\, dt$$
$$\leq \int_{t_1}^{t_2} \frac{1}{a(t)}\, dt,$$

with equality for radial light signals.

So the set of fundamental observers that can send signals at time t_1 that *eventually* reach points on γ_0 is contained in a ball in Σ of radius

$$R(t_1) := \int_{t_1}^{\infty} \frac{1}{a(t)}\, dt$$

(where the upper limit of integration should be replaced by the least upper bound of I, if one exists). Equivalently: the events in Σ_{t_1} that can send a signal that will eventually reach γ_0 form a ball of radius $a(t_1) \cdot R(t_1)$.[46]

EXAMPLE 8.4. Minkowksi spacetime is a Euclidean Robertson-Walker manifold with $I = \mathbb{R}$ and $a(t) \equiv 1$. So, relative to any choice of canonical coordinates, we find that for any time t_1, $R(t_1)$ diverges. So $I^-(\gamma_0)$ includes all of Σ_{t_1}. Now we can rehearse a familiar argument: since t_1 was arbitrary, $I^-(\gamma_0)$ includes all of Minkowski spacetime; so if (M',g') is a Lorentz manifold observationally indistinguishable from Minkowski spacetime, it must include a region isometric to Minkowksi spacetime; but since Minkowski spacetime is inextendible (being complete), this means that (M',g') must be isometric to Minkowski spacetime. More generally: if (M,g) is an inextendible Euclidean Robertson-Walker manifold and $R(t) = \infty$ for all $t \in I$, then any Lorentz manifold observationally indistinguishable from (M,g) is isometric to (M,g).

EXAMPLE 8.5. Next let us consider a toy-model version of a standard cosmological model.[47] Let (M,g) be a Euclidean Robertson-Walker manifold with $I = (0,\infty)$ and with its scale factor a satisfying the following condition: there exist $t_1, t_2 \in I$ with $t_1 < t_2$ such that $a(t) = \sqrt{t}$ for $t < t_1$ and $a(t) = e^t$ for

---

[46] All of this can be readily adapted to other Robertson-Walker manifolds—for the simply connected case, see O'Neill, *Semi-Riemannian Geometry*, corollary 12.25.

[47] What follows is a simplified version of the treatment of Ringström, *Topology and Future Stability*, §6.3.

$t > t_2$ (so that this toy model has the asymptotics of a standard cosmological model).

Then for any $t \in I$ we have:

$$R(t) \leq \int_0^\infty \frac{1}{a(t)} \, dt$$

$$= \int_0^{t_1} t^{-1/2} \, dt + \int_{t_1}^{t_2} \frac{1}{a(t)} \, dt + \int_{t_2}^\infty e^{-t} \, dt.$$

Each integral on the right hand side is finite. So there is a uniform bound on $R(t)$: we can find B so that $R(t) < B$ for all $t \in I$.

This allows us to construct a flat Robertson-Walker manifold with compact Cauchy surfaces that is observationally indistinguishable from (M, g). We proceed as follows. Choose a closed cube C in $\mathbb{E}^3$ (the space of fundamental observers) of side length $2B$ (so that C is large enough to contain an inscribed sphere of radius B). Let $M^* = I \times C$. This is a subset of M, so g induces a metric g^* on M^*. For each $t \in I$, let $C_t \subset M^*$ be the intersection of M^* with the Cauchy surface Σ_t of M: each C_t is a cubical subregion of Euclidean space with side length $2B \cdot a(t)$. Construct $\overline{M}$ by beginning with M^* and then, for each $t \in I$, identifying points opposite to each other on the boundary of C_t. Let $\bar{g}$ be the Lorentz metric that g^* induces on $\overline{M}$. Then $(\overline{M}, \bar{g})$ is a Robertson-Walker manifold whose distinguished Cauchy surfaces are flat three-tori.

Fix canonical coordinates on (M, g) and consider the fundamental observer γ_0 at rest at the spatial origin of these coordinates. The chronological past of γ_0 within (M, g) is contained in the interior of M^*—so there is a fundamental observer in $(\overline{M}, \bar{g})$ whose chronological past is isometric to that of γ_0. Indeed, in (M, g) any two fundamental observers are related by an isometry and so have isometric chronological pasts, and likewise in $(\overline{M}, \bar{g})$, since its space of fundamental observers is homogeneous. So the chronological past of any fundamental observer in (M, g) is isometric to the chronological past of some fundamental observer in $(\overline{M}, \bar{g})$, and vice versa. Further, because of the behaviour of the scale factors of (M, g) and $(\overline{M}, \bar{g})$ in the $t \to \infty$ limit, both solutions admit future conformal completions with spacelike boundaries.[48] So, as in de Sitter spacetime, the chronological

---

[48] For the case of (M, g), see, e.g., Ashtekar, Bonga, and Kesavan, "Asymptotics with a Positive Cosmological Constant. I," §3.5. The conformal boundary of $(\overline{M}, \bar{g})$ will be a quotient of the conformal boundary of (M, g).

past of any eternal observer in (M, g) or $(\overline{M}, \bar{g})$ will be identical with the chronological past of some fundamental observer with the same endpoint on $\mathcal{I}^+$. So (M, g) and $(\overline{M}, \bar{g})$ are observationally indistinguishable.

EXAMPLE 8.6. Above we saw that, under mild assumptions, if (M, g) is a Euclidean Robertson-Walker manifold such that $R(t)$ is infinite for all $t \in I$, then any Lorentz manifold observationally indistinguishable from (M, g) had to be isometric to it, while if if $R(t)$ was finite and bounded as a function of t, then we could find a Robertson-Walker manifold with toroidal spatial slices that is observationally indistinguishable from (M, g). We now look at the intermediate case, in which $R(t)$ is finite for all t, but grows without bound as t tends towards the left endpoint of I.[49]

Consider the unit de Sitter hyperboloid dS in four spacetime dimensions (the set of points lying one unit of spacelike distance from the origin in five-dimensional Minkowski spacetime, equipped with the Lorentz metric induced by the ambient Minkowski metric). Let γ_0 be a timelike geodesic in dS and restrict the de Sitter metric to the cosmological patch of γ_0 (= the chronological future of this geodesic). The resulting Lorentz manifold is extendible and geodesically incomplete—since it is a proper part of de Sitter spacetime). It is also a Euclidean Robertson-Walker manifold whose metric can be written:

$$ds^2 = -dt^2 + e^{2t} \left[dx^2 + dy^2 + dz^2 \right],$$

where $t, x, y, z \in \mathbb{R}$ and the worldline of γ_0 is given by $x = y = z = 0$. In this case, for any $t_0 \in \mathbb{R}$, we have

$$R(t_0) = \int_{t_0}^{\infty} e^{-t}\, dt$$
$$= e^{-t_0}.$$

So the intersection of $I^-(\gamma_0)$ with any Σ_{t_0} is a finite ball determines a finite ball in the space of fundamental observers—but the volume of this ball diverges as $t_0 \to -\infty$ (see Figure 8.1).

---

[49] This example is discussed by Glymour: see "Topology, Cosmology, and Convention," 211 f.; and Glymour, "Indistinguishable Space-Times," 59. See also Ringström, *Topology and Future Stability*, §7.6.

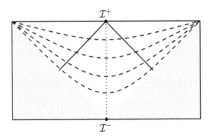

Figure 8.1 The light region is the cosmological patch of the observer represented by the dotted line. The dashed lines are level surfaces of cosmological time. The solid lines indicate the boundary of the chronological past of the observer.

Suppose that we follow the recipe of the preceding example, giving it as input now the cosmological patch and an arbitrary cube $C \subset \mathbb{E}^3$. Then the spatially toroidal Robertson-Walker manifold that we construct will not be observationally indistinguishable from the cosmological patch: in the spatially toroidal geometry, sufficiently early Cauchy surfaces will be contained in the chronological pasts of fundamental observers; this will never happen in the cosmological patch; so eternal observers will be able to determine that they live in a spacetime with non-trivial spatial topology.[50]

But we can proceed as follows: choose $t_0 \in \mathbb{R}$ and choose a cube C in the space of fundamental observers large enough to strictly contain a ball of radius e^{-t_0}. Then if we proceed as in Example 8.5, we end up with an Robertson-Walker manifold with flat three-torus Cauchy surfaces. This spatially finite Robertson-Walker manifold is not observationally indistinguishable from the cosmological patch (since eternal observers in the spatially compact geometry have complete early Cauchy surfaces in their chronological pasts). But if we delete the regions $t \leq t_0$ from both geometries, the resulting flat Robertson-Walker manifolds *are* observationally indistinguishable.

This sort of example is the motivation for the following:

---

[50] Let (M, g) be a globally hyperbolic spacetime with compact Cauchy surfaces. Under what conditions does there exist an observer in (M, g) whose chronological past includes a complete Cauchy surface? It suffices that (M, g) does *not* include an inextendible and achronal null geodesic—see Lesourd, "Observations and Predictions from Past Lightcones," lemma 3.8.

DEFINITION 8.7 (Glymour). Globally hyperbolic Lorentz manifolds (M_1, g_1) and (M_2, g_2) containing Cauchy surfaces Σ_1 and Σ_2 are (Σ_1, Σ_2)-*observationally indistinguishable* if the Lorentz manifolds that result from excising events up to and including Σ_1 in M_1 and Σ_2 in M_2 are observationally indistinguishable.

Glymour doesn't provide much motivation for this notion. But in there is in fact a compelling reason (which Glymour may well have had in mind) for being interested in *certain* pairs of cosmological models that admit observationally indistinguishable truncations:

> Since one has reason to believe that at early times in the universe's history it was filled with a hot dense ionized gas, there is for every observer a surface S in space-time (say, the surface on which the optical depth reaches unity) which is a surface, of 'last scattering' of light; he cannot obtain direct information about earlier times by optical or radio observations, since light emitted at earlier times will have been multiply scattered or absorbed by the intervening plasma.[51]

Very roughly speaking: during the first period of its existence, our Universe was too hot and too dense for stable atoms to form or for photons to travel significant distances without collisions. But as it expanded it cooled and became less dense. After about 300,000 years, when it was only 1,000 times more dense than it is today and its temperature had fallen to about 3,000 degrees Kelvin (roughly the surface temperature of a red giant like Betelgeuse), stable atoms could form and the Universe became more or less transparent to photons. On one reasonable idealization of observers, it makes sense, when thinking about more or less realistic models of universes like our own, to restrict attention to the portion subsequent to the surface of last scattering. So it sometimes makes good sense to ask about whether certain of their truncations are observationally indistinguishable.

EXERCISE 8.14 (Easier). Let (M, g) be a Euclidean Robertson-Walker manifold. Let $t_1 \in I$ and suppose that $R(t_1) = \infty$.

a) Show that $R(t) = \infty$ for all $t < t_1$.
b) Show that $R(t) = \infty$ for all $t > t_1$.

---

[51] Ellis, "Topology and Cosmology," 17. This paper is cited by Glymour in "Indistinguishable Space-Times."

EXERCISE 8.15 (Medium). Set up a Euclidean Robertson-Walker manifold with $I = (0, \infty)$ that is also a Friedmann-Lemaître solution with $\Lambda > 0$ and non-vanishing dust and radiation fluids. Show that there is B such that $R(t) < B$ for all $t \in I$.[52]

EXERCISE 8.16 (Medium). Let D be a de Sitter spacetime and let $\bar{D}$ be the corresponding elliptic de Sitter spacetime. Show that D and $\bar{D}$ are not observationally indistinguishable. Define a modestly generalized notion of observational indistinguishability and show that D and $\bar{D}$ are observationally indistinguishable in this generalized sense (aim to stay as true as possible to the spirit of the original sense).

EXERCISE 8.17 (Medium). Show that if a complete Lorentz manifold is observationally indistinguishable from a d-dimensional de Sitter spacetime ($d \geq 3$), then it is isometric to that de Sitter spacetime.

EXERCISE 8.18 (Harder). Fix $\Lambda_0 > 0$. Let $M = \mathbb{R} \times S^3$, and consider the Λ_0-vacuum Einstein equations. Friedrich's result on the global non-linear stability of de Sitter spacetime tells us that an initial data set on S^3 sufficiently similar to trivial de Sitter initial data determines a maximal globally hyperbolic development that is weakly asymptotically de Sitter and null geodesically complete.[53] Let (M, g) be a solution that arises in this way. Let (N, h) be a complete (and hence inextendible) Λ_0-vacuum solution that is observationally indistinguishable from (M, g). Show that (N, h) and (M, g) are isometric.[54]

5. Topology of Space Forms

We have seen that Euclidean Robertson-Walker manifolds may be observationally indistinguishable in one or another sense from a flat but spatially compact Robertson-Walker manifold. So for this class of geometries, observation sometimes underdetermines topology. But: the constructions that we have used show at most that we are free to substitute one topology consistent

---

[52] HINT: consult §§6.3 and 28.2 f. of Ringström, *Topology and Future Stability*.

[53] See, e.g., Valiente Kroon, *Conformal Methods in General Relativity*, chapter 15.

[54] HINT: use Beem, Ehrlich, and Easley, *Global Lorentzian Geometry*, proposition 6.16; Galloway, "Some Global Results for Asymptotically Simple Space-Times," theorem 2.1; and Gao and Wald, "Theorems of Gravitational Time Delay and Related Issues," corollary 1.

with spatial flatness for another such topology. How restrictive, topologically speaking, is the decision to use a (flat) Robertson-Walker geometry to model the Universe—and hence to model space by a (flat) space form?

Recall, first, that every space form can be constructed from a hyperbolic, Euclidean, or spherical space by quotienting out by a suitable family of isometries (the construction is sketched in Remark 8.1). Here are a couple of familiar examples.

EXAMPLE 8.7. The elliptic plane is the quotient of the two-sphere by the two-element group of isometries consisting of the antipodal map and the identity map.

EXAMPLE 8.8. The quotient of n-dimensional Euclidean by the group generated by unit translations in n linearly independent directions is a flat n-torus.

We want to investigate the sense in which space forms are topologically special among nice manifolds and the sense in which flat space forms are topologically special among space forms—our primary interest will of course be in the three-dimensional case, but we will also mention some facts about other cases. It will often be convenient to restrict attention to compact manifolds: it is a basic fact of life that the compact case is typically much better-understood—e.g., because in any dimension $n \geq 2$, there are countably infinitely many topologies for smooth compact manifolds and uncountably many topologies for smooth non-compact manifolds.[55]

For the purposes of cosmology, restriction of attention to the compact case is in any case very natural. Indeed, among mathematicians, it is often taken for granted that cosmological models should be spatially compact (so that spatial slices with the structure of full hyperbolic or Euclidean space are ruled out). Many mathematical questions about general relativity are naturally approached via consideration of the space of initial data for the Einstein equations (as are many questions of central physical interest, such as questions of global non-linear stability, cosmic censorship, and so on). Almost always, in order to gain traction on such questions, one needs to do more than simply fix a topology Σ for spatial slices and consider the set of all Riemannian metrics h and second fundamental forms k defined on Σ—one

[55] See Eichhorn, *Global Analysis on Open Manifolds*, 504. Recall that every smooth manifold admits a Riemannian metric—see, e.g., O'Neill, *Semi-Riemannian Geometry*, lemma 5.25.

needs to also assume that the relevant space of initial data has, at the every least, a topology suited to the analytical techniques that are to be brought to bear. Typically, this involves requiring that certain integrals over Σ of functions of h, k, and their derivatives converge. This will be immediate if Σ is compact or if h and k satisfy suitable fall-off conditions at spatial infinity (as in the case of initial data suited to model isolated systems in the $\Lambda = 0$ setting). But the requisite convergence will not occur when Σ is non-compact and h and k are non-trivial even at spatial infinity (as will be the case, e.g., for perturbations of Robertson-Walker type initial data).[56]

REMARK 8.1 (Quotient Geometries). Let M be a manifold and let Γ be a group of diffeomorphisms of M. The *orbit* of a point $x \in M$ is the set:

$$\Gamma \cdot x := \{g(x) \,|\, g \in \Gamma\}.$$

The *quotient* of M by Γ is the set of orbits,

$$M/\Gamma := \{\Gamma \cdot x \,|\, x \in M\}.$$

We henceforth suppose that Γ is countable. We say that Γ *acts freely* if for each $x \in M$, the identity element is the only member of Γ that fixes x. We say that Γ *acts properly* if for any $x, y \in M$, there are neighbourhoods U and V of x and y such that the set:

$$\{g \in \Gamma \,|\, U \cap g(V) \neq \varnothing\}$$

is finite. We are going to be interested in the case where Γ acts freely and properly on M.[57]

If Γ acts freely and properly on M, then M/Γ carries a natural manifold structure: under the topology according to which a subset U of M/Γ is

---

[56] On this point, see Ringström, *The Cauchy Problem in General Relativity*, §17.2. See also the discussion of Bartnik and Isenberg, "The Constraint Equations," §3; and Ringström, *Topology and Future Stability*, chapter 2.

[57] There are many common variants on this terminology. According to definition 7.6 of O'Neill, *Semi-Riemannian Geometry*, for instance, Γ is *properly discontinuous (and acts freely)* if: (i) each $x \in M$ has an open neighbourhood U such that $d(U) \cap U = \varnothing$ for all non-identity $d \in \Gamma$; and (ii) if $x, y \in M$ are not related by any element of Γ, then they have open neighbourhoods U and V such that for every $d \in \Gamma$, $d(U) \cap V = \varnothing$. This condition is equivalent to Γ's acting freely and properly in the present sense—see Lee, *Manifolds and Differential Geometry*, proposition 1.107.

open if and only if the set of points in M that compose the orbits making up U is an open subset of M, M/Γ is locally homeomorphic to M.[58] And if M carries a Lorentz or Riemannian metric of which each element of Γ is an isometry, then M/Γ inherits from M a metric of the same type that renders the two manifolds locally isometric.[59] It is a fundamental fact that every complete Riemannian or Lorentz manifold of constant curvature k is a quotient (by a countable group of isometries acting freely and properly) of a simply connected space with constant curvature k.[60]

Here is a concrete form of the construction in the Riemannian case. Fix $n \geq 2$. Let X be $\mathbb{H}_n$, $\mathbb{E}_n$, or S_n. Let Γ be a countable subgroup of the isometry group of X. If Γ act freely and discontinuously on X, then the distance function $\bar{d} \colon X/\Gamma \times X/\Gamma \to [0, \infty)$ defined by:

$$\bar{d}(\Gamma \cdot x, \Gamma \cdot y) = \inf\{d(x', y') \mid x' \in \Gamma \cdot x, y' \in \Gamma \cdot y\}.$$

gives X/Γ the structure of space form with the same curvature as X.[61]

5.1 Space Forms Are Topologically Special

In dimension two, the Uniformization Theorem tells us that every compact manifold admits a Riemannian metric of constant curvature.[62] So in this setting, space forms are not topologically special—they are, rather, the only game in town.

The situation is quite different in three dimensions: there are compact three-manifolds that cannot carry Riemannian metrics of constant curvature (see Exercise 8.19 below).

The three-dimensional analog of the Uniformization Theorem is the Geometrization Conjecture (due to Thurston, subsequently established by Perelman). It states, roughly, that each compact and oriented three manifold

[58] See, e.g., O'Neill, *Semi-Riemannian Geometry*, proposition 7.7; or Lee, *Manifolds and Differential Geometry*, proposition 1.108.

[59] See, e.g., O'Neill, *Semi-Riemannian Geometry*, corollary 7.12.

[60] See, e.g., Wolf, *Spaces of Constant Curvature*, theorem 2.4.9. In the Riemannian case, the simply connected spaces of constant curvature are the hyperbolic, Euclidean, and spherical spaces. In the Lorentz case, they are the anti-de Sitter, Minkowski, and de Sitter spacetimes.

[61] See, e.g., Ratcliffe, *Foundations of Hyperbolic Manifolds*, theorems 6.6.1, 8.1.3, and 8.5.2.

[62] See, e.g., Lee, *Introduction to Riemannian Manifolds*, theorem 3.22. Note that although $\mathbb{R}^n$ can be given a hyperbolic metric or a Euclidean metric, no compact manifold admits a metric of constant sectional curvature k for two distinct values of k from $\{-1, 0, 1\}$—see, e.g., Benedetti and Petronio, *Lectures on Hyperbolic Geometry*, proposition C.5.10.

can be decomposed into pieces, each of which either admits a metric of constant curvature (in which case this metric is of course locally homogeneous and locally isotropic) or admits a metric that locally looks like one of five other model geometries (in which case the metric is locally homogeneous but not locally isotropic).[63] So there is a natural sense in which the topologies of compact and oriented three-manifolds that admit a metric of constant curvature are quite special.[64]

EXERCISE 8.19 (Medium). Find a compact three-manifold that does not admit a Riemannian metric of constant curvature.[65]

5.2   Flat Space Forms Are Topologically Special

In both both the two- and the three-dimensional case, flat space forms are topologically special among space forms: there are infinitely many topologies for space forms, finitely many for flat space forms.

Consider, first, the two-dimensional case. There are just two topologies for surfaces of constant positive curvature: the sphere and the elliptic plane.[66] Five topologies are possible for two-dimensional flat space forms: those of the plane, the cylinder, the Möbius strip, the torus, and the Klein bottle.[67] Meanwhile, there are infinitely many topologies for two-dimensional hyperbolic space forms (indeed, for each $k \geq 2$, a sphere with k handles added can be given a metric of constant negative curvature).[68]

The picture is a bit more complex in the three-dimensional case. There are again infinitely many topologies for hyperbolic space forms in three

---

[63] For an excellent overview of the Geometrization Conjecture and its proof, see Morgan, "Recent Progress on the Poincaré Conjecture and the Classification of 3-Manifolds." Note that no compact three-manifold admits metrics g_1 and g_2, with g_1 locally isometric to one of the eight Thurston geometries and g_2 locally isometric to a different Thurston geometry—see, e.g., theorem 5.2 of Scott, "The Geometries of 3-Manifolds" (an excellent pre-Perelman overview of the Geometrization Conjecture).

[64] At the same time, there is also a natural sense in which *typical* compact and oriented three-manifolds admit metrics of constant negative curvature—see, e.g., Benedetti and Petronio, *Lectures on Hyperbolic Geometry*, chapter E.

[65] HINT: if a compact Riemannian manifold K admits a metric of constant curvature k, then so does its simply connected universal cover $\widetilde{K}$.

[66] Indeed, for any even $n \geq 2$, the only n-dimensional space forms of constant positive curvature are spherical and elliptic n-space. See, e.g., Ratcliffe, *Foundations of Hyperbolic Manifolds*, theorem 8.2.3.

[67] See, e.g., Wolf, *Spaces of Constant Curvature*, theorem 2.5.5.

[68] See, e.g., Ratcliffe, *Foundations of Hyperbolic Manifolds*, §9.7.

dimensions.[69] There are, again, only finitely many topologies for flat space forms: eight non-compact topologies (four orientable, four non-orientable), ten compact (six orientable, four non-orientable).[70] But now there are countably infinitely many topologies for spherical space forms.[71]

REMARK 8.2 (A Sense in which Flat Space Forms Are Numerous). We should also note, however, that there is a sense in which flat space forms are overwhelmingly more numerous than spherical or hyperbolic space forms. Let us specialize to the compact case. In any dimension, fixing the topology of a spherical space form fixes its metric up to isometry.[72] This is certainly not the case of compact flat space forms.[73] If, for example, you make a torus by identifying points on opposite sides of a rectangle, the geometry that you end up with depends in various ways on the side-lengths of the rectangle that you start with. So, in any dimension $n \geq 2$ there are only countably many geometrically distinct metrics for spherical space forms, while there are uncountably many for flat space forms.

What about the hyperbolic case? Here things are clearest if we also impose orientability. In dimension $n \geq 3$, compact orientable hyperbolic space forms exhibit the same sort of metric rigidity as spherical space forms: specifying the topology of a compact orientable hyperbolic space form determines its metric up to isometry.[74] So in three dimensions, there are, up to isometry, countably many spherical or hyperbolic orientable and compact space forms and uncountably many flat orientable and compact space forms. The two-dimensional case is different: if a compact orientable surface S admits a hyperbolic metric, then it admits an uncountable family of geometrically

---

[69] See, e.g., Benedetti and Petronio, *Lectures on Hyperbolic Geometry*, §§E.3 f.

[70] See Wolf, *Spaces of Constant Curvature*, §3.5. For aid in visualizing these options, see Conway, Burgiel, and Goodman-Strauss, *The Symmetries of Things*, chapter 24. Note that in any dimension $n \geq 2$, there are only finitely many topologies for compact flat space forms, each of which admits a torus as a finite cover—see, e.g., Wolf, *Spaces of Constant Curvature*, theorem 3.3.1.

[71] This is true in all odd dimensions $d \geq 3$. See Wolf, *Spaces of Constant Curvature*, chapter 7. Note that each such space is orientable—see, e.g., Ratcliffe, *Foundations of Hyperbolic Manifolds*, theorem 8.2.4.

[72] See de Rham, "Complexes à automorphismes et homéomorphie différentiable," §8.

[73] For a parameterization of flat metrics for the two-torus and the Klein bottle up to isometry, see Wolf, *Spaces of Constant Curvature*, propositions 2.5.9 and 2.5.11. For the parameterization of these families of metrics up to similarity, see, e.g., Ratcliffe, *Foundations of Hyperbolic Manifolds*, §9.5.

[74] This is a consequence of the Mostow Rigidity Theorem—see, e.g., Ratcliffe, *Foundations of Hyperbolic Manifolds*, corollary 11.8.1.

distinct hyperbolic metrics.[75] But, remarkably, each of these will assign S the same total volume.[76] In the flat case, by contrast, there are continuum-many possible volumes for a two-torus. So in two dimensions, every compact orientable spherical space form has the same volume, there are countably many possible volumes for compact orientable hyperbolic space forms—and uncountably many possible volumes for compact orientable flat space forms.

6. Ringström on Approximate Observational Indistinguishability

A powerful result due to Ringström highlights an important sense in which evidence can be expected to underdetermine cosmic topology in the $\Lambda > 0$ regime.[77] His approach is encapsulated in a remark concerning the role of Friedmann-Lemaître solutions in cosmology.

Note that the justification for using them is based not only on observations, but also on the philosophical idea that all observers should see something which is roughly similar (an assumption which cannot be tested). In practice, the assumption that leads to the standard models is that every observer sees exactly the same spatially homogeneous and isotropic solution. Clearly, this is asking too much, since what we see is not exactly spatially homogeneous and isotropic. An assumption which would be slightly more reasonable would be to fix a standard model and to say that every observer should see something which is very close to that standard model. It is of interest to ask what limitations on the topology such an assumption imposes; note that the standard perspective, which implies a locally homogeneous and isotropic spatial geometry, is only consistent with a topology which is the 3-sphere, hyperbolic space or Euclidean space, or a quotient thereof.[78]

[75] See, e.g., Ratcliffe, *Foundations of Hyperbolic Manifolds*, corollary 9.7.1.
[76] This follows from the Gauss-Bonnet Theorem—see, e.g., Ratcliffe, *Foundations of Hyperbolic Manifolds*, theorem 9.3.1.
[77] For a summary, see Ringström, "On the Future Stability of Cosmological Solutions to Einstein's Equations with Accelerated Expansion." For details, see his *Topology and Future Stability*. For related results, see his "Future Stability of the Einstein-Non-Linear Scalar Field System."
[78] Ringström, "On the Future Stability," 997.

That is: to require local isotropy is to restrict attention to space forms—a strong topological restriction in three spatial dimensions. Ringström goes on to argue that if we replace the assumption of local isotropy by an assumption of *approximate* local isotropy, then there is a sense in which it is "not possible to draw any conclusions concerning the topology of the universe" from our observations.[79] We begin by filling in some relevant background to Ringström's approach.

6.1 Background

6.1.1 The Einstein-Vlasov Equations

First, a brief introduction (omitting mathematical detail) to the Einstein-Vlasov equations, which provide the analytic setting for Ringström's results.[80]

The Vlasov equation (also known, more justly, as the *collisionless Boltzmann equation*) is an evolution equation native to the kinetic theory of gases. Recall the first version of statistical mechanics pioneered by Maxwell and by Boltzmann, today usually called *kinetic theory*. A complete micro-description X of a gas of N molecules assigns a precise position and momentum to each molecular constituent of the gas (for convenience, we assume each constituent molecule has the same mass). Given such a precise microstate, we can give a sort of rough summary of its content as follows: break the one-particle phase space $\Gamma_1 = \{(q,p) \mid q, p \in \mathbb{R}^3\}$ into a countable infinity of little cells (each with the same finite volume); count, for each of these cells, how many molecules have position and momentum falling within it, according to X. In this way, our exact microstate of the gas determines a *histogram function* h_X on Γ_1: for any $(q,p) \in \Gamma_1$, $h_X(q,p)$ is the number of molecules that (according to X) occupy the unique cell containing (q,p). The histogram h_X corresponding to a microstate X encodes a great deal of information about X.[81] But a histogram is hard to work with, since it is only piecewise continuous. So in kinetic theory, one works instead with a

---

[79] Ringström, "On the Future Stability," 997.

[80] For thorough introductions, see, e.g., Choquet-Bruhat, *Introduction*, chapter X; or Rendall, *Partial Differential Equations in General Relativity*, §3.4.

[81] But it also leaves a lot out. Even in the $N = 1$ case, it replaces an exact microstate with coarse-grained information about the position and momentum of the particle. And in the multi-particle case, in constructing the histogram, we discard all information concerning correlations between particles—on the crucial importance of such information, see Batterman, *A Middle Way*.

smoothed-out surrogate f_X for h_X. If we normalize f_X then we end up with a probability distribution over the one-particle phase space: $N^{-1}f_X(q,p)$ gives us a good estimate of the probability density for a randomly chosen particle to have position near q and momentum near p according to X.

So far we have discussed how instantaneous states of many-body systems are represented in kinetic theory. The Vlasov equation is a partial differential equation for a time-dependent scalar field $f(q,p,t)$ on Γ_1 that provides a simple-minded account of the dynamics of a smoothed-out version of a histogram. Assume that collisions between constituents of the gas can be ignored. If no forces act on the gas molecules, then each of them moves inertially. And this induces a dynamics on the instantaneous states $f(q,x)$ of kinetic theory: in this simplest of cases, the Vlasov equation just amounts to requiring that $f(q,p)$ be constant along inertial trajectories.[82] If, say, we wish to treat the gas as a self-gravitating system, then we ask how a molecule with position q and momentum p would move under the influence of the gravitational potential determined by the smoothed out mass distribution $\rho(q) = \int f_X(q,p)\,dp$, and this again determines an evolution law for the density $f(q,p)$.[83]

Of course, since this approach neglects collisions, it is radically inadequate to modelling typical samples of gas—since it is collisions (and other close encounters) between molecules that drive the rapid equilibration of such systems. So the Vlasov equation provides a useful dynamical model of a gas only on extremely short timescales. But the same basic approach can be used to model astrophysical systems.[84] Loosely speaking, a galaxy can be treated as a gas of stars and a super-cluster can be treated as a gas of galaxies (again, everything is simplest if we treat all constituents as having equal mass). And in the astrophysical setting, collisions and close encounters play only a very small role in the dynamics of large systems. So in this setting, the timescales on which the Vlasov equation provides a good approximation are enormous (longer than the age of the Universe, for typical galaxies).[85]

---

[82] Fix a position in space x_0 and a momentum p. Let $x(t)$ be the position at time t of an inertially moving particle initial with initial position x_0 and momentum p. Then the Vlasov equation dictates that $f(x(t),p,t)$ is a constant function of t.

[83] In this case, the Vlasov equation tells us that f is constant along any trajectory of a particle moving in the external gravitational field ρ determined by f (where f and ρ may be time-dependent).

[84] On the history of this approach, see Walter, "Mathematical Milky Way Models from Kelvin and Kapteyn to Poincaré, Jeans and Einstein."

[85] On this point, see Binney and Tremaine, *Galactic Dynamics*, §7.1.

This gives you the gist of the physical content of the Vlasov equation.[86] Of course, there is also a relativistic Vlasov equation (where now $f_X(x, v)$ needs to be a function that at each spacetime point x determines a distribution over the relevant mass hyperboloid in the tangent space at that point)—and this can be coupled to the Einstein equation to yield the Einstein-Vlasov equations. For present purposes, it is important that although the Einstein-dust equations and the Einstein-Vlasov equations both allow one to model large collections of massive bodies, models employing the latter are less liable to develop unphysical singularities—in particular, shock-type shell-crossing singularities are endemic in dust models (even in flat spacetime, with gravity turned off) but are absent in solutions of the Einstein-Vlasov equations.[87] Further, in the setting of expanding cosmological models, it is possible to choose initial data so that the Vlasov matter mimics a radiation fluid at early times and mimics dust at late times.[88]

6.1.2 The Power of Local Assumptions when $\Lambda > 0$

Let (M, g) be a globally hyperbolic Lorentz geometry. Let Σ be a Cauchy surface in M and let B be a metric ball in Σ (relative to the Riemannian metric that g induces on Σ). Recall that the *causal future* of B, denoted $J^+(B)$, is the set of points that can be reached from B by future-directed causal curves while the *future domain of dependence* of B, denoted $D^+(B)$, is the set of points p such that every past inextendible causal curve through p intersects B. Intuitively, if (M, g) models a world in which physical influences can travel no faster than the speed of light, then $J^+(B)$ is the region that can be influenced by events in B while $D^+(B)$ is the region whose state is fully determined by events in B.

Here are two Euclidean Robertson-Walker metrics we could place on $\mathbb{R}^4$:

$$ds_1^2 = -dt^2 + dx^2 + dy^2 + dz^2$$
$$ds_2^2 = -dt^2 + e^{2t}\left[dx^2 + dy^2 + dz^2\right].$$

---

[86] For full details and many astrophysical applications, see Binney and Tremaine, *Galactic Dynamics*.

[87] For discussion and references, see Rendall, "The Nature of Spacetime Singularities," §4; and Rendall, "The Einstein-Vlasov System," §5. A positive cosmological constant appears to have at least some tendency to counteract shock formation—see Mondal, "The Nonlinear Stability of $n + 1$-Dimensional FLRW Spacetimes."

[88] See Ringström, *Topology and Future Stability*, §1.3 and chapter 28. As Ringström notes, it would in some ways be more natural in approximating standard cosmological models to employ two species of Vlasov matter (one massive, one massless)—see his remark 7.35 and §10.2.

The first gives $\mathbb{R}^4$ the structure of Minkowski spacetime, the second the structure of a de Sitter cosmological patch. As usual, for $t \in \mathbb{R}$, we write Σ_t for $\{(t, x, y, z) \mid x, y, z \in \mathbb{R}\}$—which will inherit a Euclidean structure from either of these Lorentz metrics. And for any $x \in \mathbb{R}^3$ we write γ_x for the curve $\gamma_x : t \to (t, x)$ (which corresponds to a fundamental observer, according to either of the Lorentz metrics above). As usual, γ_0 stands for the fundamental observer at rest at the origin of the spatial coordinates.

For any $r > 0$, we use B_r to denote the sphere of radius r in Σ_0 centred at the origin. Note that for the two metrics mentioned above, the scale factor for Σ_0 is unity—so we can identify Σ_0 with the space of fundamental observers.

In Minkowski spacetime, for any $r > 0$, $D^+(B_r)$ will be a portion of a lightcone (see Figure 8.2). Only the dimensions can vary: the apex of $D^+(B_r)$ will lie r units of proper time above Σ_0. So, in particular, for any finite fixed r, $D^+(B_r)$ cannot contain arbitrarily long timelike geodesic segments.

Things are quite different in the de Sitter cosmological patch (see Figure 8.3).[89] The analysis of Example 8.6 above shows that $D^+(B_1)$ includes each point p on γ_0 to the future of Σ_0: for if p is such a point and γ is an inextendible past-directed causal curve departing from p, then γ must intersect Σ_0 in some point q (since Σ_0 is a Cauchy surface); but then q must lie in B_1 (since B_1 is the set of points on Σ_0 that can send signals to points on γ_0). Further, this analysis shows that for any $r < 1$, $D^+(B_r)$ will include only a finite segment of γ_0 to the future of Σ_0—from which it follows that γ_0 is the only worldline of a fundamental observer whose portion to the future of Σ_0 is contained in $D^+(B_1)$.

What does $D^+(B_2)$ look like? Well, the same reasoning that establishes that the entire future portion of γ_0 lies in $D^+(B_1)$ establishes that the entire future portion of any fundamental observer γ_x lies in the future domain of dependence of the ball of radius one centred at x. So, in particular, for each

Figure 8.2 The qualitative behaviour of $D^+(B_r)$ in Minkowski spacetime for $r = \frac{1}{2}$, $r = 1$, and $r = 2$ (scaled so that the bases all have the same width).

---

[89] Here we follow Ringström, *Topology and Future Stability*, §7.6.

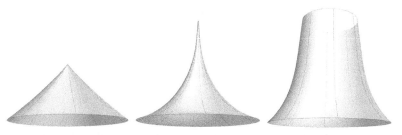

Figure 8.3 The qualitative behaviour of $D^+(B_r)$ in the cosmological patch of de Sitter spacetime. for $r = \frac{1}{2}$, $r = 1$, and $r = 2$ (scaled so that the bases all have the same width). $D^+(B_{\frac{1}{2}})$ shows only a slight deformation from its Minkowski counterpart. The other two are of necessity truncated, as both are infinite in vertical extent. $D^+(B_2)$ includes the future portions of all fundamental observers in B_1, $D^+(B_1)$ just the future portion of γ_0.

$x \in B_1$, the entire future portion of γ_x is contained in $D^+(B_2)$ (since every point in B_1 is the centre of a ball of radius one contained in B_2). (You are asked to continue this analysis of the cosmological patch in the next two exercises below.)

We find that in the cosmological patch, we can control the state within large chunks extending to arbitrarily late times by controlling the state within a suitable spatial ball at any time. The relevance of all of this is that a similar picture holds very generally for solutions of the $\Lambda > 0$ Einstein-Vlasov equations—a fact that plays a key role in the proof of Ringström's result.[90]

EXERCISE 8.20 (Easier). Continue the analysis of the cosmological patch begun above: show that $J^+(B_1)$ is entirely contained in $D^+(B_3)$.

EXERCISE 8.21 (Easier). Still using the metric for the cosmological patch, consider any time $t' \in \mathbb{R}$ and any point x' in Σ'_t. Write $\gamma_{x'}$ for the fundamental observer who passes through x' and write B'_r for the Euclidean ball in $\Sigma_{t'}$ of radius r centred at x' (using the metric on $\Sigma_{t'}$ induced by the spacetime metric). Show that there are r_1, r_2, and r_3 such that: $D^+(B'_{r_1})$ contains the portion of $\gamma_{x'}$ to the future of $\Sigma_{t'}$ (and contains no other infinite portions of worldlines determined by fundamental observers); $D^+(B'_{r_2})$ includes the future portions of every fundamental observer that intersects B'_{r_1}; and $J^+(B'_{r_1})$ is contained in $D^+(B'_{r_3})$.

---

[90] See his discussion in *Topology and Future Stability*, §§7.6 f.

6.2 Ringström's Result

Ringström distinguishes between the Copernican Principle ("there are no privileged observers") and the Cosmological Principle ("the Universe is spatially homogeneous and isotropic").[91] His starting point, in the line of thought we are following, is the observation that although the Cosmological Principle implies very strong constraints on spatial topology (only special compact oriented three-manifold topologies are consistent with Riemannian metrics of constant curvature, almost none with flat metrics), what we really want to know is what sort of spatial topologies are consistent with an approximate version of that principle. Ringström offers the following informal gloss on the relevant result:

> given a background solution [i.e., a suitable standard cosmological model], a $t_0 \ldots$, a closed 3-manifold Σ, and an $\varepsilon > 0$, there is a solution to the Einstein-Vlasov equations such that the spatial topology is Σ and such that, in regions that can be seen by causal observers, the solution is closer to the background than ε. When we say 'can be seen,' we take for granted that observations concern the future of the $t = t_0$ hypersurface...., Due to the existence of these solutions, it is fair to say that the assumption that every observer considers the Universe to be almost spatial homogeneous and isotropic does not impose any restrictions on the spatial topology. Moreover, it seems hard to argue that the above solutions violate the Copernican principle. On the other hand, disregarding a negligible collection of topologies, they are necessarily incompatible with the cosmological principle.[92]

Here is a slightly more detailed sketch of (an implication of) the result. Consider a standard cosmological model (M, g) with non-trivial dust and radiation content and cosmological constant $\Lambda > 0$. Let Σ be a canonical Cauchy surface in (M, g), let $\ell \in \mathbb{N}$, let $\varepsilon > 0$, and let K be any compact and oriented three-manifold. Then there is a solution (N, h) of the Λ-Einstein-Vlasov equations such that:

a) N is of the form $J \times K$ for some open interval $J \subset \mathbb{R}$.

---

[91] Ringström, *Topology and Future Stability*, 6 f.

[92] Ringström, *Topology and Future Stability*, 79. The relevant result is stated in §7.9 of that work and proved in §34.4. As Ringström emphasizes, his full result implies that the picture developed is non-linearly stable in an interesting sense.

b) (N, h) is globally hyperbolic and future timelike and null geodesically complete.

c) There is a Cauchy surface Σ' for (N, g) such that the portion of (M, g) to the future of Σ and the portion of (N, h) to the future of Σ' are approximately observationally indistinguishable in the sense that for any observer in one of these portions, we can find an observer in the other portion, such that the causal pasts of these observers (within their respective portions) come within ε of being isometric to one another (relative to some technical criteria built out of C^ℓ norms).[93]

In brief: for any standard cosmological model, any instant of time Σ in that model, and any any compact and oriented three-manifold topology, there is a solution of the Einstein-Vlasov equations with the given spatial topology and an instant of time Σ' in that solution, such that the two models are approximately (Σ, Σ')-observationally equivalent. Crucially, if we take Σ_1 to represent the surface of last scattering then, as Ringström suggests in his gloss above, it is reasonable to think of the portions of these models to the future of the given instants of time as modelling the observable portions of the Universe.

Note that we cannot expect a result of this kind if we require strict (Σ, Σ')-observational equivalence rather than the approximate variety: if a Robertson-Walker manifold is (Σ, Σ')-observationally equivalent to a standard cosmological solution, then its Cauchy surfaces must be flat—but, of course, we know that almost no compact and oriented three-manifolds admit flat Riemannian metrics.

6.3 Coda

Ringström also briefly discusses an intriguing conjecture (due to Anderson and to Fischer and Moncrief) according to which even in the vacuum $\Lambda = 0$ regime, the Einstein equations typically drive expanding universes towards states in which most parts of space are approximately locally isotropic.[94]

---

[93] See §7.9 of Ringström, *Topology and Future Stability* for a fully detailed formulation of the result (including, of course, the relevant sense of ε-closeness).

[94] See Ringström, *Topology and Future Stability*, §3.3. The conjecture derives from Anderson, "On Long-Time Evolution in General Relativity and Geometrization of 3-Manifolds"; and Fischer and Moncrief, "The Reduced Einstein Equations and the Conformal Volume Collapse of 3-Manifolds." For an assessment of the $\Lambda > 0$ and non-vacuum regimes, see Moncrief and

The picture is roughly as follows. We are interested in globally hyperbolic vacuum solutions that begin with a singularity and which expand eternally towards the future. We restrict attention to solutions with compact Cauchy surfaces of a topology Σ inconsistent with a Riemannian metric of non-negative scalar curvature (this is a weak restriction on compact topologies). We assume that our solutions can be foliated by Cauchy surfaces of constant mean curvature, with this curvature tending towards 0 from below as $t \to \infty$. Relative to this foliation, we can think of the Einstein equations as governing the evolution in time of the geometry of space—this gives us the *Einstein flow* in the space of Riemannian metrics on Σ.[95]

It turns out that hyperbolic metrics play a special role in this dynamics: e.g., the Einstein flow on Σ has a fixed point if and only if Σ admits a hyperbolic metric. One form of the conjecture that we are interested in is: any admissible Σ can be decomposed into pieces H that admit hyperbolic metrics and pieces S that do not; under the conditions specified above, as $t \to \infty$ the volume of the hyperbolizable pieces will grow far more rapidly than the volume of the other pieces—and observers located in the hyperbolizable regions will see the Universe as being approximately locally isotropic at late times. If this conjecture can be established, it would give us a sort of cosmic no-hair result that does not require $\Lambda > 0$.

QUESTION 8.4 (Smaller?). Here is Ringström's commentary on the significance of this conjecture.

If one believes the universe to be close to spatially homogeneous and isotropic, one question naturally arises: is this due to the evolution, i.e., a consequence of the Einstein equations, or is it due to our universe being very special? In the above division of Σ into H and S, the suggestion that the metric on H should converge to a hyperbolic metric indicates that there is isotropisation on H. However, there is no isotropisation on S. On the other hand, the volume of S is, asymptotically, negligible relative to the volume of H. In other words, the fraction of observers considering the universe to tend to become isotropic tends to unity asymptotically.

Mondal, "Could the Universe Have an Exotic Topology?"; and Moncrief and Mondal, "Einstein Flow with Matter Sources."

[95] For a brief introduction to this approach, see §7.3 of Belot, "The Representation of Time and Change in Mechanics."

Should this form of isotropisation be considered to be consistent with the Copernican principle?[96]

Well, should it? Would this conjecture, if established, deliver a satisfying sense in which the approximate isotropy that we observe is just what we should expect to see in a general relativistic world—even if $\Lambda = 0$?

<hr>

[96] Ringström, *Topology and Future Stability*, 42.

9
Brains!

1. Introduction

Some philosophers love skeptical scenarios. Maybe most of our beliefs are false because we are merely dreaming (Zhuangzi). Or, though awake, we are about to enter a state that stands to wakefulness as wakefulness stands to dreaming (al-Ghazali). Or we are deceived by an evil demon (Descartes). Or we and the world around us came into being only five minutes ago (Russell).

Of course, in their working lives, scientists tend to have little time for such conceits—geologists and palaeontologists do not pause to worry about the possibility that the world around us came into being five minutes ago or 5,000 years ago, replete with misleading traces of a much longer history. And for this reason, naturalistically minded philosophers tend to be impatient with standard skeptical scenarios, which rely on elements—demons, brain-vats, varieties of dreaming very different from human dreaming—that arise only in possible worlds remote from the actual world (and thus as being ignorable in assessment of workaday knowledge claims).[1]

Here we will be concerned with the problem of Boltzmann brains. This is a skeptical worry reminiscent of Hume: perhaps much of the apparently lawful structure of appearances that has obtained so far was the result of brute luck—and we are in for some nasty surprises shortly. We will, as usual, be asked to reflect on the situation of epistemically unlucky counterparts of ourselves—but with the twist that we will be given reasons deriving from physics for thinking that we have many such counterparts in the actual world. Unlike the classic skeptical scenarios, this one is taken seriously by many working scientists. Responses proposed in the physics literature include (amongst others): that the observed exponential expansion of the Universe cannot be driven by a true cosmological constant; that we should assign prior probability zero to any cosmological theory that leads to this problem; that we need to reconceive confirmation in cosmology so that what is confirmed is a package consisting of a cosmological theory and a

---

[1] For this sort of stance, see Maddy, *What Do Philosophers Do?*

Accelerating Expansion: Philosophy and Physics with a Positive Cosmological Constant. Gordon Belot, Oxford University Press. © Gordon Belot 2023. DOI: 10.1093/oso/9780192866462.003.0010

self-locating probability distribution; that we need to consider modifying Born's rule for calculating probabilities in quantum theories; and that it is a flaw in a theory to predict infinite lifespans for spatially finite universes.[2]

So the problem of Boltzmann brains offers something for everyone—a skeptical scenario that even naturalists should love. Here is the version of the argument that we will focus on.

i) According to currently favoured cosmological theories, there should exist many objects that are (essentially) physical duplicates of you and your current environment.
ii) Among these near-duplicates of you will be some that are thinking beings with evidence and beliefs.
iii) The vast majority of these are parts of causal chains much shorter than than those we usually take our lives to be part of: they inhabit relatively short-lived fluctuations out of thermal soup.
iv) So, compared to what we normally take our epistemic situation to be, the vast majority of thinking near-duplicates of you have a great deal of misleading evidence and a great many false beliefs about the past and future.
v) So unless you have reason to believe that you are special—i.e., not typical among thinking beings in our world that are (essentially) perfect physical duplicates of you—you should expect that much of your evidence is misleading and that many of your beliefs about the past and future are false.

Note that the argument does not require that all of your near-duplicates have the same beliefs as you do—just that at least some that inhabit relatively short-lived fluctuations have some sort of beliefs. So adopting some form of externalism about content (or about life) will not on its own defuse the argument (on this point, see Remark 9.2 below). Similarly, the argument does not suppose that you and those of your near-duplicates that have beliefs have the same evidence or the same justifications. So adopting some form of epistemological externalism will not on its own defuse the argument.[3]

[2] See Dyson, Kleban, and Susskind, "Disturbing Implications of a Cosmological Constant," 17; Carroll, "Why Boltzmann Brains are Bad," §6; Srednicki and Hartle, "Science in a Very Large Universe," §VII.A; Page, "Possibilities for Probabilities," 2; Page, "Is Our Universe Likely to Decay within 20 Billion Years?"

[3] For an influential defence of externalism about evidence, see Williamson, *Knowledge and Its Limits*, chapters 8 and 9. On Williamson's approach, there are many propositions that are known

There are, of course, ways to evade the argument's unpleasant conclusion. We could argue that currently favoured cosmological theories should be replaced (as noted above, this is the moral drawn by some cosmologists). Or we could argue that intelligent near-duplicates of us are too rare for the argument to get off the ground. Or we could argue that, surprisingly enough, our intelligent near-duplicates typically have just as many true beliefs as we take ourselves to have. Or we could provide some reason for thinking that we are special rather than typical—or that typicality, properly understood, should not in this case lead us to revise our assessment of our epistemic situation.

Although we will make a few remarks as we go about other aspects, our primary interest here is in the first step of this argument. In Section 2, we review a few basic facts about statistical mechanics that will drive the argument. Section 3 is devoted to the early roots of the problem of Boltzmann brains in writings of Boltzmann and Eddington. In Section 4 we discuss Eddington's reasons for concluding that the problem of Boltzmann brains does not arise in future asymptotically de Sitter cosmologies. In Section 5 we survey the reasons that modern cosmologists have given for reversing this verdict. Section 6 collects a number of questions about this most perplexing topic.

REMARK 9.1 (Nomenclature). As we will see below, the problem of Boltzmann brains has a long history—but not, until relatively recently, under that name.[4] The modern resurgence of interest in the problem dates to a paper of 2002 that argued that the problem afflicts the cosmological models dynamically dominated by Λ at late times.[5] The problem only got its catchy name a couple of years later.[6] By Kripkean baptism, a Boltzmann brain is a:

> brain (complete with 'memories' of the Hubble deep fields, microwave background data, etc.) fluctuating briefly our of chaos and then immediately equilibrating back into chaos again.[7]

by an agent situated in an ordinary environment but which are merely believed by duplicates of that agent situated in the sort of aberrant environment that typical skeptical scenarios involve—and on his view, a proposition is part of an agent's evidence if and only if it is known by the agent.

[4] For a helpful overview and references, see Albrecht, "Tuning, Ergodicity, Equilibrium, and Cosmology."

[5] Dyson, Kleban, and Susskind, "Disturbing Implications." See also chapter 4 of Dyson, *Three Lessons in Causality*.

[6] See Albrecht and Sorbo, "Can the Universe Afford Inflation?" For a while, *freak observers* vied with *Boltzmann brains* for terminological supremacy.

[7] Albrecht and Sorbo, "Can the Universe Afford Inflation?," §III.C.

When we talk about objects in longer-lived fluctuations, we could speak, as some authors do, of Boltzmann People, Boltzmann Species, Boltzmann Planets, Boltzmann Galaxies, and so on.[8] I will avoid this: as the reader will shrewdly suspect, philosophers tend to be more worried than physicists are about what it takes for an object to be a brain, a person, or a species (see Remark 9.2 below). As I aim to remain neutral on as many such questions as possible, my primary use of the phrase *Boltzmann brain* below will be as part of my official name for the problem we are focusing on, rather than for a type of object.

2. Statistical Physics—A Quick Review

The starting point of statistical physics is the observation that although it is often bootless to ask what a many-body system is certain to do—e.g., because the equations of motion for such systems are typically impossibly difficult to solve explicitly—it is nonetheless often profitable to ask what is overwhelmingly likely to happen to such a system.

Recall, first, the basic apparatus of Boltzmann-style statistical mechanics.[9] Consider a system made up of N point-particles moving in a region of space R (depending on the application, each point-particle might correspond to a gas molecule, a star, or a galaxy). We use Γ_1 to denote the one-particle phase space, points of which assign a momentum and a position in R to a single particle. We use Γ to denote the full $6N$-dimensional phase space of the system, points of which assign a momentum and a position in R to each of the N particles. We assume that the dynamics is given by a time-independent Hamiltonian on Γ of the usual kinetic-plus-potential-energy form: so, in particular, specifying a state $z_0 \in \Gamma$ at time $t = 0$ determines a state $z(t)$ at any time t.[10] For each way E of fixing the conserved quantities of the system, we use Γ_E to denote the subset of Γ consisting of states in which the conserved quantities of the system are given by E.[11] We call points in Γ

---

[8] See, e.g., Carroll, "Why Boltzmann Brains are Bad," §3.

[9] For a brief overview, see, e.g., Goldstein and Lebowitz, "On the (Boltzmann) Entropy of Non-Equilibrium Systems."

[10] At this point we have tacitly deleted from Γ any states that determine singular dynamical trajectories—to see how this works for the paradigmatic case of hard spheres in a box, see Cercignani, Illner, and Pulvirenti, *The Mathematical Theory of Dilute Gases*, appendix 4.A.

[11] Depending on the context, a way E of fixing the values of the conserved quantities either involves specifying a precise value for each such quantity or specifying, for each such quantity,

or in a Γ_E *fine-grained states.* Because energy is always among the conserved quantities (and in many applications it is the only conserved quantity, or nearly the only one), we will call each Γ_E an *energy surface.*

The phase space Γ comes equipped with a natural measure μ (the Liouville measure), which induces a measure μ_E on each Γ_E.[12] We will denote the measure of a subset S of Γ or Γ_E by $|S|$, which we will call the *volume* of S.

Relative to any partition $\mathcal{M}$ of the phase space Γ (or of an energy surface Γ_E) into subsets M_1, M_2, ... of positive finite volume, we can assign an entropy to each fine-grained state z as follows: the *Boltzmann entropy of z relative to $\mathcal{M}$* is:

$$S_B(z) := \ln|M(z)|,$$

where $M(z)$ denotes the unique element of $\mathcal{M}$ to which z belongs.

In what follows, we will focus on physically salient $\mathcal{M} = \{M_j\}$ with the feature that two states in the same M_i are macroscopically indistinguishable (or nearly so).[13] We call the elements of such a partition $\mathcal{M}$ *macro-states.*

For present purposes, there is a crucial divide between systems whose energy surfaces have finite volume and systems whose energy surfaces have infinite volume. We consider each case in turn.

2.1 Systems with Finite-Volume Energy Surfaces

Intuitively: the problem of Boltzmann brains is generated when we consider an eternal system in which (typical) dynamical trajectories move 'at random' wandering from one macro-state to another, with the volume of any given macro-state measuring the proportion of time that the dynamical trajectory spends in it.

---

an extended interval in which its value lies, chosen from an antecedently specified family of intervals that partition the set of possible values of the quantity.

[12] See, e.g., Abraham and Marsden, *Foundations of Mechanics*, §3.4.

[13] In many applications, we can follow Boltzmann in using a partition $\mathcal{M}$ of the following form: choose some way to break the one-particle phase space Γ_1 into countably many cells $c_1, c_2, \ldots$ of equal volume; then decree that two fine-grained states z_1 and z_2 are equivalent if they agree, for each cell c_j of Γ_1, about how many molecules have states in c_j; then take $\mathcal{M}$ to consist of the corresponding equivalence classes. Another common method is to fix a set of observables of interest, partition the ranges of each of them into small cells, and then let each macro-state correspond to a choice of cell for each of the observables.

In many cases of interest, the energy surfaces Γ_E have finite volume.[14] Typically, given a finite-volume energy surface Γ_E for a many-particle system, it will be possible to select a physically reasonable family of macro-states that includes a distinguished macro-state M_0 that eats up almost all the volume of Γ_E.[15] In this case we call M_0 the state of *thermodynamic equilibrium*. For a system like a diffuse gas in a box, this will be a state in which the gas particles are roughly evenly distributed in space (with their momenta roughly satisfying a Gaussian distribution).

Every system with finite-volume energy surfaces exhibits Poincaré recurrence—but this on its own does not generate the problem Boltzmann brains. Some systems with finite-volume energy surfaces exhibit a stronger behaviour, ergodicity, which is sufficient to generate the problem of Boltzmann brains. But we don't expect actual systems with finite-volume energy surfaces to be ergodic. We do, however, expect those with many degrees of freedom to exhibit a sort of approximately ergodic behaviour. Let us consider each of these points in turn.

POINCARÉ RECURRENCE. Let U be an open subset of Γ_E, let $z_0 \in U$, and let $z(t)$ be the dynamical trajectory determined by $z(0) := z_0$. We say that z_0 is *recurrent* in U under our dynamics if we can find a sequence of times $t_1, t_2, \ldots$ with $t_k \to \infty$ such that $z(t_k) \in U$ for each k. The *Poincaré Recurrence Theorem* tells us that if Γ_E has finite volume, then the set of $z \in U$ that are *not* recurrent in U is a set of μ_E-measure zero.[16]

---

[14] This will happen, for instance, if: (i) the system of interest has finitely many degrees of freedom; (ii) the region of space R accessible to it is compact; and (iii) the potential energy of the system is bounded from below. For in this case, possible values of the position's degrees of freedom will be bounded (being a subset of R^N for some N) and so too will be the possible values of the momenta (given that the total energy is E and that kinetic energy is non-negative, the only way for the set of possible momenta to be unbounded is if there is no lower bound on the potential energy). Examples of systems without finite-volume energy surfaces: (a) a perfect fluid confined to a compact region of space satisfies (ii) and (iii) but not (i) (see Arnold and Khesin, *Topological Methods in Hydrodynamics*, §II.4.B); (b) a finite gas of hard spheres free to roam through all of Euclidean space (without gravity) satisfies (i) and (iii) but not (ii) (see Section 2.2 below for discussion and references); (c) a finite system of point-particles confined to a box but interacting gravitationally satisfies (i) and (ii) but not (iii) (see, e.g., Lynden-Bell and Wood, "The Gravo-Thermal Catastrophe in Isothermal Spheres and the Onset of Red-Giant Structure for Stellar Systems").

[15] On this point see, e.g., Goldstein and Lebowitz, "Entropy of Non-Equilibrium Systems," §3. For some exceptions to this rule of thumb, see Goldstein *et al.*, "Macroscopic and Microscopic Thermal Equilibrium," §8.

[16] See, e.g., Oxtoby, *Measure and Category*, chapter 17; or Coudène, *Ergodic Theory and Dynamical Systems*, §1.3. For discussion of some related results, see Wallace, "Recurrence Theorems." Poincaré recurrence is not generic for systems whose energy surfaces have infinite volume such as Examples (a)–(c) of fn. 14 above.

So (measure-zero exceptions aside) the dynamical trajectory determined by initial data z_0 comes arbitrarily close to z_0 infinitely often in the future. This is surprising:

> For instance, if one were to burn a piece of paper in a closed system, then there exist arbitrarily small perturbations of the initial conditions such that, if one waits long enough, the piece of paper will eventually reassemble (modulo arbitrarily small error)![17]

However, on its own, Poincaré recurrence is consistent with the dynamics of being periodic (see the discussion of harmonic oscillators below). And periodic behaviour is not enough to give us the problem of Boltzmann brains: a Universe could be periodic in time without any of its inhabitants being in the grip of surprisingly many false beliefs about the past and the future.

ERGODICITY. In this discussion, we assume that E corresponds to an assignment of precise values to the conserved quantities of our system.

For any measurable $U \subseteq \Gamma_E$ and any $z_0 \in \Gamma_E$, the *mean sojourn time* of z_0 in U is the limit as $t \to \infty$ of the proportion of $[0, t]$ spent in U by the dynamical trajectory $z(t)$ passing through z_0 at $t = 0$. We call our dynamics on Γ_E *ergodic* if for any measurable $U \subseteq \Gamma_E$ and for μ_E-almost all z_0, the mean sojourn time of z_0 in U is $\mu_E(U)/\mu_E(\Gamma_E)$. We call our dynamics on Γ *ergodic* if the dynamics on Γ_E is ergodic for each E.[18] Ergodic systems exhibit two striking forms of behaviour (here typical = aside from a set of exceptional cases of μ_E-measure zero).

> EQUILIBRATION: In an ergodic system with a state of thermodynamic equilibrium M_0, typical dynamical trajectories spend the overwhelming majority of their time in that state (since M_0 eats up almost all of the volume of the energy surface).

> FLUCTUATIONS: Every macro-state will contain an open subset of Γ_E. Each open region $U \subset \Gamma_E$ is visited infinitely often by dynamical trajectories determined by typical initial data—so such dynamical trajectories are dense in Γ_E and visit each macro-state infinitely often.

---

[17] Tao, *Poincaré's Legacies*, remark 2.8.4.
[18] For a brief historical overview of ergodic theory, see Moore, "Ergodic Theorem, Ergodic Theory, and Statistical Mechanics."

There are important examples of ergodic classical mechanical systems. Consider the most basic model of a dilute gas: N hard spheres in a box. Let $N \geq 2$ spheres of radius r move in a cubical region R of Euclidean space (where r is small relative to R). Each sphere moves inertially unless its centre of motion comes within $2r$ of the centre of another sphere or r of the boundary of R, in which case an elastic collision occurs. It is believed that this system is ergodic for all $N \geq 2$—but to date this has only been established in the $N = 2$ case.[19] More attention has been focused on the related and mathematically more tractable model in which R is endowed with periodic boundary conditions (so that the particles live on a torus), which is known to be ergodic for all $N \geq 2$.[20]

There are also natural examples of non-ergodic classical mechanical systems with finite-volume energy surfaces. Consider a harmonic oscillator in the plane.[21] The phase space Γ for this system is four-dimensional. A generic energy surface Γ_E is determined by fixing the values h and ℓ of the energy and the angular momentum with $|\ell| < h$. Each such Γ_E is a two-torus—and the Poincaré recurrence theorem applies. But the dynamics is periodic—each dynamical trajectory wraps around Γ_E just once (and so will not visit every macro-state—unless we have set things up in a very unnatural way). So, in particular, no dynamical trajectory is dense in Γ_E. Recurrence does not imply ergodicity.

SETTLING FOR SECOND BEST. It is thought to be unlikely that many realistic classical models of physical systems are ergodic: indeed, there is a sense in which generic classical mechanical systems with finite-volume energy surfaces are *not* ergodic.[22] But according to a widely accepted picture, amongst realistic classical models with finite-volume energy surfaces, those which model systems with large numbers of constituents can be expected to exhibit a form of behaviour nearly as impressive:

EFFECTIVE ERGODICITY. As measured by μ_E, the vast majority of points $z_0 \in \Gamma_E$ have the following feature: as $t \to \infty$, the proportion of time that the dynamical trajectory $z(t)$ determined by z_0 spends in any macro-state

[19] See Simányi, "Ergodicity of Hard Spheres in a Box."

[20] For discussion and references, see Szász, "Boltzmann's Ergodic Hypothesis"; and Simányi, "Further Developments of Sinai's Ideas."

[21] For a detailed analysis, see, e.g., Cushman and Bates, *Global Aspects of Classical Integrable Systems*, chapter I.

[22] See Markus and Meyer, *Generic Hamiltonian Dynamical Systems Are Neither Integrable Nor Ergodic*.

M is approximately the proportion of points in Γ_E that lie in M (relative to certain physically reasonable ways of choosing a family of macro-states).[23]

Effective ergodicity is weaker than the real thing in two ways: the behaviour is posited for the overwhelming majority of micro-states, rather than for all but a measure-zero set of exceptional states; and the claim covers only macro-states, not arbitrary open sets.

We expect a many-particle effectively ergodic many-body system to feature a macro-state M_0, the state of thermodynamic equilibrium which eats up almost all of the volume of Γ_E. And we expect that, for such a system, aside from some special exceptions, each dynamical trajectory will spend almost all of its time in the state of thermodynamic equilibrium—but will visit lower-entropy macro-states infinitely often, with the proportion of time it spends in a given macro-state being measured by its volume (so that higher-entropy macro-states are exponentially favoured over lower-entropy macro-states). The cash value of effective ergodicity is: confidence that we will not go too far wrong if we expect that, with respect to common macro-properties, well-behaved large systems should exhibit the same sort of equilibration and recurrence behaviour as ergodic systems do.

2.2 Systems with Infinite-Volume Energy Surfaces

Unless confined by forces, N point-particles moving in an unbounded region of space R will have a classical model in which energy surfaces have infinite volume. Further, even if the particles are restricted to move in a bounded region, we can have $|\Gamma_E| = \infty$ if the potential energy is not bounded below: a finite system of self-gravitating point-particles provides an example.[24]

Since we have set things up so that macro-states always have finite volume, if Γ_E has infinite volume there can be no state of thermodynamic equilibrium that eats up almost all the volume of the energy surface.[25] Systems of

---

[23] For versions this picture, see, e.g., Ruelle, "Ergodic Theory," 613 f.; Gallavotti, *Statistical Mechanics*, §9.3; Lebowitz, "From Time-Symmetric Microscopic Dynamics to Time-Asymmetric Macroscopic Behavior," 72. For further discussion and references, see Frigg and Werndl, "Explaining Thermodynamic-Like Behavior in Terms of Epsilon-Ergodicity."

[24] For details, see, e.g., Padmanabhan, "Statistical Mechanics of Gravitating Systems"; or Kiessling, "The Influence of Gravity on the Boltzmann Entropy of a Closed Universe."

[25] A confined self-gravitating system will have finite-volume energy surfaces and macro-states of thermodynamic equilibrium if its constituent particles are given a finite size (as in the model of a dilute gas discussed above). But of course the state of thermodynamic equilibrium

gas molecules or stars free to explore all of Euclidean space fall under this case—such systems do not approach a state of thermodynamic equilibrium because there is no such thing.

> In order that a mass of gas can be in thermodynamic equilibrium, it is necessary that it be enclosed. There is no thermodynamic equilibrium of a (finite) mass of gas in an infinite space.[26]

Einstein made the point in terms of a self-gravitating gas in thermal equilibrium with temperature T and a gravitational potential Φ that goes to zero at spatial infinity (the boundary conditions appropriate for a finite self-gravitating system).[27] Boltzmann's formula describing the density of a system in equilibrium in a potential Φ applies:

$$\frac{\rho(x_1)}{\rho(x_0)} = e^{-(\Phi(x_1)-\Phi(x_0))/T}.$$

Let x_0 be a point at which ρ and Φ are both non-zero. By choosing x_1 far enough away, we can then make the left hand side of Boltzmann's formula as small as we like—but the right hand side will approach $e^{\Phi(x_0)/T} > 0$ as x_1 tends to spatial infinity. This is a contradiction—so no such point x_0 can exist and so such a system cannot be in a state of thermodynamic equilibrium.

At the level of statistical mechanics: if we have a finite self-gravitating system, then we expect that from time to time (through statistical fluctuations) enough kinetic energy will be acquired by one of the components for it to achieve escape velocity and to never see its friends again—so such a system will slowly evaporate until just two gravitationally bound constituents remain.[28] So for an isolated N-body system free to explore all of Euclidean space, we do not expect to see an approach to equilibrium, effectively ergodic behaviour, or even recurrence.[29]

---

will then look very different from that of a dilute gas—in particular, it will have the particles tightly clustered rather than being evenly distributed throughout the container. See, e.g., Padmanabhan, "Statistical Mechanics of Gravitating Systems."

[26] Gibbs, *Elementary Principles in Statistical Mechanics*, 35 n.

[27] Einstein, "Cosmological Considerations in the General Theory of Relativity," §1 (6.43).

[28] See again §1 of Einstein, "Cosmological Considerations" (6.43). For a modern analysis of this process, see Binney and Tremaine, *Galactic Dynamics*, §7.5.2.

[29] This does not automatically carry over for arbitrary situations in infinite spaces. Recall that Einstein's argument above covered the case in which $\Phi \to 0$ as $r \to \infty$. If instead $\Phi \to \infty$ as $r \to \infty$, then no particle can escape to spatial infinity, no matter what its initial velocity. The singular isothermal sphere is a workhorse astrophysical toy model whose gravitational

3. Boltzmann and Eddington Get Us into a Mess

Let us get down to business. The essential pieces of a version of the problem of Boltzmann brains can be found already in Boltzmann and Eddington.

Suppose that you unearth a long-buried box containing a diffuse gas. You can be morally certain that it will be in a state of thermodynamic equilibrium. So, for example, if you select two different regions of the box in which to perform measurements of temperature (or density, or pressure, or ...) you can count on registering outcomes essentially indiscernible from one another. At the same time, you can also be morally certain that if you took enough such pairs of measurements, you would occasionally register a pair of markedly differing outcomes, as a result of statistical fluctuations. Minuscule fluctuations happen all the time, appreciable ones very rarely. Large ones happen unimaginably rarely. You can expect to wait longer than the age of the Universe before finding all of the molecules in one-half of the box.

Already in 1895, Boltzmann was interested in a cosmological version of this scenario, in which we picture the Universe as a whole as a vast but finite gas-in-a-box system in which we expect there to exist occasional pockets of organization, which Boltzmann calls *worlds*.

> I will conclude this paper with an idea of my old assistant, Dr. Schuetz.
>
> We assume that the whole universe is, and rests for ever, in thermal equilibrium. The probability that one (only one) part of the universe is in a certain state, is the smaller the further this state is from thermal equilibrium; but this probability is greater, the greater is the universe itself. If we assume the universe great enough, we can make the probability of one relatively small part being in any given state (however far from the state of thermal equilibrium), as great as we please. We can also make the probability great that, though the whole universe is in thermal equilibrium, our world is in its present state. It may be said that the world is so far from thermal equilibrium that we cannot imagine the improbability of such a state. But can we imagine, on the other side, how small a part of the whole universe this world is? Assuming the universe great enough, the probability that such a small part of it as our world should be in its present state, is no longer small.

potential exhibits this behaviour—see, e.g., Binney and Tremaine, *Galactic Dynamics*, §4.3.3.b. For a general relativistic analog, see Reiris, "Stationary Solutions and Asymptotic Flatness. I," §1.2.

If this assumption were correct, our world would return more and more to thermal equilibrium; but because the whole universe is so great, it might be probable that at some future time some other world might deviate as far from thermal equilibrium as our world does at present. Then the afore-mentioned H-curve would form a representation of what takes place in the universe. The summits of the curve would represent the worlds where visible motion and life exist.[30]

(Boltzmann's quantity H is the negative of entropy—for his rendering of the H-curve, see Figure 9.1).

Let us call a cosmological model a *Boltzmann universe* if is temporally infinite in both directions, involves a finite number of bodies moving in a finite box (with reflecting or periodic boundary conditions), and its thermo-dynamic behaviour is analogous to a box of gas (so there is a macro-state of thermodynamic equilibrium and the dynamics is effectively ergodic in the sense discussed above).

Boltzmann was himself somewhat ambivalent about the scientific cre-dentials of cosmological speculation.[31] Thirty-five years later, Eddington had seen relativistic cosmology develop into an observational science and had no such qualms. He drew attention to some striking features of Boltz-mann universes in his Presidential Address to the Mathematical Association (formerly the Association for the Improvement of Geometrical Teaching).[32] Since fluctuations of every scale should occasionally occur,

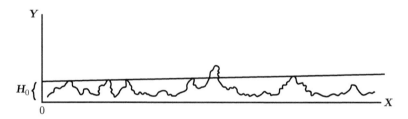

Figure 9.1 Boltzmann's rendering of the H-curve (from his reply to Zermelo).

[30] Boltzmann, "On Certain Questions of the Theory of Gases," 415. See also Boltzmann, "On Zermelo's Paper 'On the Mechanical Explanation of Irreversible Processes,'" §4. On Ignaz Schütz (1867–1927), see Darrigol, *Atoms, Mechanics, and Probability*, 376.
[31] See, e.g., Boltzmann, "On Zermelo's Paper," §4. For further discussion and references, see Darrigol, *Atoms, Mechanics, and Probability*, 314, 400, and 455.
[32] Eddington, "The End of the World." See also the related discussion in chapter III of Eddington, *New Pathways in Science* (based on his Messenger Lectures of 1934).

If we wait long enough, a number of atoms will, just by chance, arrange themselves in systems as they are at present arranged in this room; and, just by chance, the same sound-waves will come from one of these systems of atoms as are at present emerging from my lips; they will strike the ears of other systems of atoms, arranged just by chance to resemble you, and in the same stages of attention or somnolence. This mock Mathematical Association meeting must be repeated many times over—an infinite number of times, in fact—before t reaches $+\infty$.[33]

Here Eddington is speaking just a bit loosely: there is no reason to think that a *perfect* duplicate of the meeting he addressed will ever occur in the future. Rather, what we expect is that there will be events in which atoms are arranged as similarly as you like to how they were arranged during his lecture. Since smaller fluctuations occur exponentially more often than larger ones, Eddington continues,

There is no limit to the amount of the fluctuation, and if we wait long enough we shall come across a big fluctuation which will take the world as far from thermodynamical equilibrium as it is at the present moment. If we wait for an enormously longer time, during which this huge fluctuation is repeated untold numbers of times, there will occur a still larger fluctuation which will take the world as far from thermodynamical equilibrium as it was one second ago.

Let us call any event that we are willing to consider a near-perfect physical duplicate of state of the lecture hall and its contents during Eddington's address a *meeting of the Mathematical Association.* Let us call such an event a *genuine* meeting if it occurs as a part of a history something like what we are told is the history of the visible Universe since the Big Bang—where what we care about here is the spatial and temporal extension of this history, how spectacularly low the initial entropy was, and the extent to which our expectations about entropy increase, global and local, are more or less borne out in this history. And let us call such an event a *mock* meeting if it falls significantly short of this (readers should set their own standards of what constitutes a significant shortfall).

If what we know about a region of spacetime is that it is the site of a meeting of the Mathematical Association, what should we expect the world

[33] Eddington, "The End of the World," 320.

to look like just before and after this event and just outside this region during the meeting? Eddington tells us that our train of thought has led us to the following:

> it is practically certain that a universe containing mathematical physicists will at any assigned date be in the state of maximum disorganisation which is not inconsistent with the existence of such creatures.[34]

Let us call that *Eddington's Principle*. An implication of this principle is that mock meetings of the Mathematical Association are astronomically more common than genuine meetings: there are many macro-states of a Boltzmann universe that include a subsystem in some given state of organization; and among these macro-states, those in which the world exterior to the given subsystem are highly disorganized, take up vastly more phase space volume (and should therefore occur vastly more frequently) than those in which the exterior is similarly highly organized. Feynman, who speaks of fluctuations rather than mock meetings, spells out what Eddington leaves implicit:

> although when we look at the stars and we look at the world we see everything is ordered, if there were a fluctuation, the prediction would be that if we looked at a place where we have not looked before, it would be disordered and a mess. Although the separation of the matter into stars which are hot and space which is cold, which we have seen, could be a fluctuation, then in places where we have not looked we would expect to find that the stars are not separated from space.[35]

Let us change the example slightly (and take some anachronistic liberties). Each day, the cosmic microwave background radiation that reaches us originated in a different part of space. So far, observation has revealed a temperature distribution that is very close to being isotropic. Attendees of meetings of the Mathematical Association will expect this pattern to persist. Those attending genuine meetings will (presumably) be vindicated in this expectation. Those attending mock meetings will be in for a surprise sooner or later (but later with an astronomically small probability). And the surprise will not just concern the cosmic microwave background radiation: Eddington's Principle implies that you should be very worried about your immediate

---

[34] Eddington, "The End of the World," 322.
[35] Feynman, *The Character of Physical Law*, 115 (based on his Messenger Lectures of 1964).

future if you ever see evidence that your worldline is approaching the edge of a region of order within a cosmos in thermodynamic equilibrium—since this is evidence that the pocket of order in which you have lived is about to be absorbed back into the cosmic soup.

Eddington gives us the pieces we require to argue that if you live in a Boltzmann universe, you face a version of the problem of Boltzmann brains outlined in Section 1 above. You should expect there to exist many near-perfect duplicates of you and your current environment. The vast majority of these near-duplicates occupy fluctuations from equilibrium much more short-lived than we take our own local environment to occupy. Plausibly, by any reasonable standards, at least some of your near-duplicates who occupy relatively short-lived fluctuations have beliefs—including, presumably, many more mistaken beliefs about the past and the future than we take ourselves to have. So the vast majority of near-duplicates of you who have beliefs have many false beliefs. So unless you have some reason to think that you should not be treated as a typical member of the class of near-duplicates of you with beliefs, you should think that you have surprisingly many false beliefs about the past and the future.

REMARK 9.2 (Externalisms). Some readers may think that one or another form of externalism provides an easy way to defuse the problem of Boltz-mann brains.

Let us begin with externalism about content. Can we evade the problem of Boltzmann brains by adopting a view on which it is possible for two things to be perfect physical duplicates while differing in the content of their thought and speech (if any)? No. But certain forms of externalism about content (and related doctrines) allow us to ameliorate the problem of Boltzmann brains to some degree.

Consider Davidson's Swampman: a duplicate of Davidson, created by chance when lightning strikes a tree—a "Boltzmann brain" for philosophers of language.[36] Davidson himself is committed to a fairly common form of externalism about content:

> what a person's words mean depends in the most basic cases on the kinds of objects and events that have caused the person to hold the words to be applicable; similarly for what the person's thoughts are about.[37]

---

[36] Introduced at Davidson, "Knowing One's Own Mind," 443 f.
[37] Davidson, "Knowing One's Own Mind," 456.

This leads him to deny that Swampman when first created is capable of thought or speech.[38] However, it is clear that for Davidson, Swampman can be expected to become a thinking being with evidence and beliefs, given a suitable course of interaction with the surrounding world.[39] It is not clear how long Davidson thinks this will take, but perhaps we can take a decade or two as a reasonable upper bound. So on Davidson's view, it seems, a duplicate of you can only count as having thoughts and beliefs if it has been interacting normally with its environment for some years—so we don't need to worry about the briefest (and overwhelmingly most common) sorts of fluctuations from equilibrium. But this is not too much solace—we don't need to worry that the world as we know it was created last Thursday (or a microsecond ago), but we still need to worry that the 1960s never happened—and that the organized world we see around us is about to be dissolved into thermal soup.

Not everyone is so generous to Swampman. Some, like Millikan, tie thought and reason to biological function, and tie biological function to evolutionary history.[40] For such philosophers, not only is Swampman not a thinking being, it is not a living organism and does not have sense organs or other body parts. Thompson arrives at the same view via a different route (he is an externalist about his notion of form-of-life):

> the thing has no ears to hear with and no head to turn; it has no brain-states, no brain to bear them, and no skull to close them in; prick it, and it does not bleed; tickle it, and it does not laugh; and so forth. It is a mere congeries of physical particles and not so much as alive.[41]

What does it take for something that is a virtually perfect duplicate of you to count as a thinking thing? On Davidson's view, it might be enough that the thing has been interacting normally with its environment for a sufficiently long time. On the sterner views of Millikan and Thompson, it is required that the thing be the causal product of a sufficiently long chain of animal-like things that have been interacting normally with their environments. It is hard to know what precisely is required here. But if our current science is correct, then the Earth has been around for about a third of the time that the Universe has existed. So even on the most stringent view of this kind, it

---

[38] Davidson, "Knowing One's Own Mind," 444.
[39] Davidson, "Knowing One's Own Mind," 456 n. 4.
[40] See, e.g., Millikan, "On Knowing the Meaning."
[41] Thompson, *Life and Action*, 60.

seems like we can say that a human-like thing counts as a thinking being if the fluctuation it is part of is sufficiently deep—even if its total duration is only a fraction of what we normally take the history of the Universe to be. This is enough to cause trouble: such a being would have the same chance that we normally take ourselves to have true beliefs about human, evolutionary, and geological history, but can be expected to have many false beliefs about the deeper past—and to be quite wrong in its expectations about the future.

4. Eddington's Solution

Eddington himself is unwilling to accept the conclusion to which he has led his audience.[42] His view is that there is something wrong with Boltzmann universes as cosmological models. His favoured cosmological models differ in two respects.[43] (1) They are finite towards the past, beginning in states of spectacularly low entropy.[44] (2) They are expanding—indeed, asymptotically de Sitter—towards the future.

Now, merely positing a low entropy initial state does not on its own help with the problem of Boltzmann brains.[45] After all, Eddington is worried about Boltzmann universes because they have the following feature:

(∗) Any highly organized local system, such as a roomful of (things physically identical to) mathematical physicists, that arises as a subsystem of a much larger highly organized system (such as human history) can also arise as the result of (relatively small-scale) fluctuations—and although throughout infinite time, a given organized local system will arise in both ways, it will arise astronomically more frequently in the latter way than in the former way.

---

[42] He says that it is quite clear that what we dubbed Eddington's Principle in Section 3 above does not apply to our case.

[43] Eddington prefers a sort of past-finite Eddington-Lemaître cosmology: beginning in a state close to a state of the Einstein static universe and asymptotic in the future to de Sitter spacetime. See his "The End of the World," 316 and 319; and his *New Pathways in Science*, 58 and 220.

[44] For canonical discussions of the virtues of cosmologies with low-entropy initial states, see Boltzmann, "On Zermelo's Paper"; Feynman, *The Character of Physical Law*, 115 f.; Feynman, Leighton, and Sands, *The Feynman Lectures on Physics*, volume I, §46.4; Albert, *Time and Chance*, chapter 4; and Albert, *After Physics*, chapters 1–3.

[45] On this point, see, e.g., Dyson, Kleban, and Susskind, "Disturbing Implications," §6.

Let us call a cosmological model a *semi-Boltzmann universe* if it is like a Boltzmann universe except that it has a beginning in which it is in a state of spectacularly low entropy. Of course, (∗) holds of semi-Boltzmann universes just as much as of Boltzmann universes. So positing a low-entropy beginning for the system does not help with the problem raised by Eddington—unless we have some principled reason to think that when we ask about our place in the Universe, we need only consider as possibilities situations that arise as part of the Universe's first long climb from its low-entropy beginning to its first attainment of equilibrium.

It is Eddington's second point that does the work. When he first mentions the possibility of mock meetings of the Mathematical Association due to statistical fluctuations for equilibrium, Eddington appends a footnote:

> I am hopeful that the doctrine of the "expanding universe" will intervene to prevent its happening.[46]

He offers more detailed remarks when he returns to this topic several years later.

> It was argued . . . that every possible configuration of atoms must repeat itself at some distant date. But that was on the assumption that the atoms will have only the same choice of configurations in the future that they have now. In an expanding space any particular congruence becomes more and more improbable. The expansion of the Universe creates new possibilities of distribution faster than the atoms can work through them, and there is no longer any likelihood of a particular distribution being repeated. If we continue shuffling a pack of cards we are bound sometime to bring them back into their standard order—but not if the conditions are that every morning one more card is added to the pack.[47]

Think of it this way. Above we saw that broad qualitative features of the future of a finite collection of stars or atoms often depends on whether or not the collection is restricted to move in a finite region: many systems restricted in

---

[46] Eddington, "The End of the World," 320.

[47] Eddington, *New Pathways in Science*, 68. The idea that the threat of recurrence can be avoided if the size of the relevant state space grows over time plays an important role in essays of Leibniz concerning the finitude of the humanly expressible. See the texts in Fichant (ed.), *De l'horizon de la doctrine humaine*. For commentary, see Rescher, "Leibniz and Issues of Eternal Recurrence"; and Forman, "Leibniz on Human Finitude, Progress, and Eternal Recurrence."

this way exhibit the same tendency towards equilibration and fluctuations as a box of gas; systems free to roam through infinite domains generally behave completely differently. A spatially compact but expanding universe is a sort of hybrid of these two cases—at any time, only a finite region is available for the system to explore, but there is no upper bound on how large the distance between two objects may grow with time. Clearly, if space is finite but expanding quickly enough, the system might as well be located in an infinite space (imagine a gas enclosed in a container whose walls recede at the speed of light).

Consider the case that Eddington has in mind: a future asymptotically de Sitter universe. If γ_0 and γ_1 are inextendible timelike geodesics, the only way that all of γ_1 can be in γ_0's causal past (and vice versa) is if γ_0 and γ_1 are asymptotic to one another towards the future. And, presumably, the only way that a group of massive bodies in such a spacetime can stay in causal contact with one another towards the future is if the form is a gravitationally bound system. But on statistical grounds we again expect that this will not happen: given sufficient time, statistical fluctuations will endow one particle after another with escape velocity, until there are only two left. In this scenario, we certainly shouldn't expect infinitely many copies of us to form spontaneously towards the future in a de Sitter-like universe. Indeed, given that even moderately large fluctuations are overwhelmingly rare, maybe we shouldn't expect *any* near-duplicates of us to form by chance before exponential expansion has its way.

5. The Worm Turns

There things rested for many years with the problem of Boltzmann brains: it was commonly accepted that in an expanding universe, one did not have to worry about recurrence, let alone effectively ergodic behaviour.[48] So what has changed? Why do many cosmologists now reject Eddington's verdict that the problem of Boltzmann brains doesn't arise in a universe that is finite towards the past and eternal and asymptotically de Sitter towards the future?

---

[48] On this point, see, e.g., Barrow and Tipler, *The Anthropic Cosmological Principle*, 176. In fact, a result of Tipler establishes a sense in which generic spatially compact cosmologies do not exhibit recurrence (even if they are not permanently expanding)—see his "General Relativity and the Eternal Return," §§III f.

The tide was turned by a famous paper of Dyson, Kleban, and Susskind.[49] The opening and closing paragraphs of the paper are as follows:

> As emphasized by Penrose many years ago, cosmology can only make sense if the world started in a state of exceptionally low entropy. The low entropy starting point is the ultimate reason that the universe has an arrow of time, without which the second law would not make sense. However, there is no universally accepted explanation of how the universe got into such a special state. In this paper we would like to sharpen the question by making two assumptions which we feel are well motivated from observation and recent theory. Far from providing a solution to the problem, we will be led to a disturbing crisis.[50]
>
> We wish to emphasize that the above conclusions appear to be the inevitable consequence of the following assumptions:
>
> * There is a fundamental cosmological constant.
> * We can apply the ideas of holography and complementarity to de Sitter space.
> * The time evolution operator is unitary, so that phase space volume is conserved.
>
> Perhaps the only reasonable conclusion is that we do not live in a world with a true cosmological constant.[51]

The key contribution of Dyson, Kleban, and Susskind in this paper is to bring a holographic perspective to bear on the analysis of a future asymptotically de Sitter universe. This leads them to treat the currently observationally favoured model of our Universe as sharing qualitative features with a (semi-)Boltzmann universe—and to conclude that it is *ipso facto* afflicted with the problem of Boltzmann brains.

According to a classic analysis of Gibbons and Hawking, there are deep analogies between the event horizon of a black hole and the cosmological

[49] Dyson, Kleban, and Susskind, "Disturbing Implications." See also chapter 4 of Dyson, *Three Lessons in Causality*.

[50] Dyson, Kleban, and Susskind, "Disturbing Implications," 1. No citation to a paper of Penrose is provided, but perhaps Dyson and her co-authors have in mind the discussion of the limited power of anthropic reasoning in Penrose, "Time-Asymmetry and Quantum Gravity."

[51] Dyson, Kleban, and Susskind. "Disturbing Implications," 17. The final sentence is absent from the corresponding passage in §4.4.6 of Dyson, *Three Lessons in Causality*—presumably it has been displaced by the conclusion in §4.6.1 that a de Sitter vacuum decays inevitably decays into a Minkowski vacuum.

horizon of an observer in de Sitter spacetime: in particular, the boundary of the static patch of a de Sitter observer can be assigned an entropy and a temperature—and a freely falling observer in a de Sitter vacuum should detect the corresponding thermal radiation.[52] Dyson, Kleban, and Susskind maintain that in de Sitter spacetime, as in the context of black holes, it is a mistake to

> build a quantum mechanics of the entire global spacetime—including regions which have no operational meaning to a given observer, because they are out of causal contact with that observer.[53]

They suggest, rather, that when doing physics in de Sitter spacetime, one ought to restrict attention to a single static patch. Then:

> the theory describes a closed isolated box bounded by the observer's horizon, and makes reference to no other region. Furthermore, as in the case of black holes, the mathematical description of this box should satisfy the conventional principles of linear unitary quantum evolution.[54]

More specifically, they suggest that:

> the physics within the [static patch] is exactly described by a dual quantum system that includes a hamiltonian H and a space of quantum states. Furthermore the state of the world within the [static patch] is [the] density matrix $\exp(-\beta H)$ with the appropriate temperature. The assumption is motivated by what we now know about other quantum gravity systems such as the AdS black hole.[55]

Given the physically reasonable assumption that the spectrum of H is discrete, the quantum analog of the Poincaré Recurrence Theorem assures us that any wave-function ψ_0 determines a solution $\psi(t)$ of the Schrödinger

---

[52] Gibbons and Hawking, "Cosmological Event Horizons, Thermodynamics, and Particle Creation." For a unified treatment of the thermal vacuum states of the static patch of de Sitter spacetime, the Rindler wedge in Minkowski spacetime, and the region exterior to the horizon in Schwarzschild spacetime, see Kay and Wald, "Theorems on the Uniqueness and Thermal Properties of Stationary, Nonsingular, Quasifree States on Spacetimes with a Bifurcate Killing Horizon."

[53] Dyson, Kleban, and Susskind, "Disturbing Implications," 2.

[54] Dyson, Kleban, and Susskind, "Disturbing Implications," 2.

[55] Dyson, Kleban, and Susskind, "Disturbing Implications," 12.

equation with the feature that for any $\varepsilon < 0$ there is a $T > 0$ such that $||\psi(T) - \psi_0|| < \varepsilon.$[56] On this basis, our authors assume that the relevant dynamics is effectively ergodic.[57]

In the final section of the paper, the consequences of this picture are unpacked. In this treatment, the static patch of a de Sitter observer is in effect a Boltzmann universe: there will be many chunks of spacetime macroscopically indistinguishable from our current context—but most of these will not lie ~ 13 billion years to the future of a Big Bang-style low entropy state but will instead reside in relatively shallow fluctuations from equilibrium. Plausibly, this reflects the claim made in the opening sentence of the paper: cosmology does not 'make sense' in this setting because the methods employed by cosmologists to reconstruct the history of the Universe are profoundly misleading far more often than they are reliable. This leads to the closing suggestion that the accelerating expansion of the Universe, discovered just a few years previously, is not due to a positive cosmological constant. The argument has, to put it mildly, had a large influence among physicists.[58] In the final section below, I will pepper readers with questions— and leave the hard work of evaluating this argument to them.

REMARK 9.3 (From Quantum Recurrence to Boltzmann Brains). As emphasized above: in the classical mechanical setting, a system can exhibit recurrence without being ergodic (integrable systems provide good examples— they are recurrent while, intuitively, being as far as possible from ergodicity); nonetheless, one expects that the sort of many-body systems that arise in practice exhibit effectively ergodic behaviour. In particular, it is generally safe to assume that in a many-body system with finite-volume energy surfaces, the overwhelming majority of dynamical trajectories spend almost all of their time in the state of thermodynamic equilibrium (equilibration) but also visit every macro-state infinitely often (fluctuations), with exponential suppression of large fluctuations.

How do things look in the quantum case? Fix a system of n quantum particles restricted to a spatial region of finite extent. The energy spectrum of the system will be discrete—so the quantum recurrence theorem applies.

[56] See Bocchieri and Loinger, "Quantum Recurrence Theorem"; and Schulman, "Note on the Quantum Recurrence Theorem."
[57] Obviously there is a gap here—for discussion, see Remark 9.3 below.
[58] See fn. 2 above for the tip of the iceberg. For a comprehensive list of references covering the first five or so years following the paper of Dyson, Kleban, and Susskind, see Page, "The Born Rule Fails in Cosmology."

How close does this get us to what we need to get the problem of Boltzmann brains up and running?

i) Here is one approach to making sense of equilibration in this setting.[59] We restrict attention to the case in which the energy spectrum of the system of interest is bounded from below. Then there will only be finitely many eigenvalues (counting multiplicities) in any finite interval of real numbers. Let us consider an interval J that is long enough to contain a large number of such eigenvalues but which is short enough that it is not feasible to make macroscopic distinctions between them (this should be possible for large n). Let $\mathcal{H}_J$ denote the Hilbert space spanned by the corresponding eigenvectors (this is our quantum analog of a classical energy surface). Because $\mathcal{H}_J$ is finite-dimensional, there is a natural equilibrium state ρ_J associated with it: the density matrix that is an equal-weight mixture of each of pure states corresponding to energy eigenvectors spanning $\mathcal{H}_J$. No pure state in $\mathcal{H}_J$ can assign each possible measurement outcome the same probability that ρ_J does. But it can be shown that there is a sense in which, for sufficiently large n, the vast majority of pure states in $\mathcal{H}_J$ do give almost the same probabilities of measurement outcomes as ρ_J does for any family of mutually exclusive macro-properties (= any decomposition of $\mathcal{H}_J$ into mutually orthogonal subspaces of sufficiently large dimension). Further, if certain technical conditions are satisfied by the Hamiltonian of the system and a given family of macro-properties, then *every* state ψ in $\mathcal{H}_J$ determines a history $\psi(t)$ with the feature that for the vast majority of t, $\psi(t)$ gives almost the same probabilities of measurement outcomes as ρ_J does for the given family of macro-properties. So we have a version of equilibration: from the macro perspective, the dynamical trajectory determined by any initial state will look like an equilibrium state at most future times.

ii) The situation is less clearcut when it comes to fluctuations. One thing that is clear is that there can be no state $\psi_0 \in \mathcal{H}_J$ that determines

---

[59] There are many possible approaches. Here we follow the approach pioneered by von Neumann in "Proof of the Ergodic Theorem and the H-Theorem in Quantum Mechanics" and further elaborated by Goldstein *et al.* in "Long-Time Behavior of Macroscopic Quantum Systems."

a dynamical trajectory $\psi(t)$ dense in $\mathcal{H}_J$ (or in the space of rays of $\mathcal{H}_J$). For suppose that $\phi_1, \phi_2, \ldots, \phi_N$ are energy eigenvectors that form a basis for $\mathcal{H}_J$ with corresponding eigenvalues $\lambda_1, \ldots, \lambda_N$. Consider an initial state $\psi_0 := \sum_{k=1}^{k=N} a_i \phi_i$. Then the state at time t will be $\psi(t) = \sum_{i=1}^{i=N} a_i e^{-i\lambda_k t} \phi_i$. So in particular, at any time the probability of positive outcome for an experiment to determine whether the system is in state ϕ_1 is $|a_1|^2$. So the dynamical trajectory $\psi(t)$ determined by the initial state ψ_0 is certainly not dense in $\mathcal{H}_J$, since it does not approach arbitrarily close to ϕ_1 (except in the degenerate case in which $|a_1|^2 = 1$; but in this case $\psi(t)$ will of course never come anywhere close to any of the other ϕ_k).[60]

iii) Suppose that $\psi_0 = \sum_{k=1}^{k=N} a_i \phi_i$ is typical in that each co-efficient $a_k \neq 0$ and that our system's Hamiltonian is typical in that no energy Eigenvalue λ_k is a rational multiple of any other. Then the dynamical trajectory $\psi(t)$ determined by ψ_0 will explore an $(N-1)$-real-dimensional torus $\mathbb{T}$ sitting inside the $(N-1)$-complex-dimensional space of quantum states (the space of rays of $\mathcal{H}_J$). Further, $\psi(t)$ will come arbitrarily close infinitely often to each point in $\mathbb{T}$ and the proportion of time that it spends in any region of $\mathbb{T}$ will be given by the ratio of the volume of that region to the volume of $\mathbb{T}$.[61]

iv) Does this get us all the way to the problem of Boltzmann brains? It is hard to know—one would need to know something about the range of states that make up the torus $\mathbb{T}$ explored by the dynamical trajectory determined by a given initial state ψ_0. One thing perhaps worth noting: if a system has $\sim 10^{20}$ classical degrees of freedom, then the torus that its quantum state explores should have something like $10^{10^{20}}$ dimensions.[62] So there is a sense in which in attempting to cook up wild scenarios, we have enormously more free parameters to play with in the quantum case than in the classical one. But in the absence of a worked-out account of what sort of possibilities are described by these quantum states, it is unclear how much this is really worth.

[60] For further discussion, see §0.4 of von Neumann, "Goldstein Proof of the Ergodic Theorem."
[61] See Brody and Hughston, "Geometric Quantum Mechanics," §§11–14; and Brody and Hughston, "Unitarity, Ergodicity, and Quantum Thermodynamics."
[62] For this estimate, see Goldstein et al., "Long-Time Behaviour," 177.

6. Questions

QUESTION 9.1 (Larger?).   What is the best way to close the gap (discussed in Remark 9.3 above) between quantum recurrence and the sort of fluctuation behaviour required to get the problem of Boltzmann brains up and running?

QUESTION 9.2 (Larger!).   Dyson, Kleban, and Susskind arrive at the problem of Boltzmann brains by: (i) restricting attention to the static patch of an observer in de Sitter spacetime; and (ii) inferring from the holographic principle that the static patch can be treated as a closed quantum-mechanical system.

There is another, quite distinct, route to the problem of Boltzmann brains.[63] Restrict attention to a static patch. Note that the effect of the cosmological constant will be to dilute matter content, so that any quantum fields should asymptotically approach a vacuum state at late times. Note that de Sitter spacetime carries a natural vacuum state, the Bunch-Davies vacuum, the restriction of which to a static patch is a thermal state (compare with the Unruh effect—the restriction of the Minkowski vacuum to a Rindler wedge is a thermal state).[64] Infer that every possible matter configuration eventually arises (infinitely often) as a vacuum fluctuation in a de Sitter-like future.

a) What are the advantages and disadvantages of this second route over the one pioneered by Dyson, Kleban, and Susskind?
b) Boddy, Carroll, and Pollack argue that this second route does not in fact lead to the problem of Boltzmann brains. On their view, while it is true that the vacuum state has non-trivial overlap with, e.g., a state in which duplicates of Team Canada and Team CCCP re-play the Summit Series, absent the right sort of decoherence, it is a mistake to think that a static patch featuring a stationary thermal state corresponds to a history in which the greatest event in sports history recurs infinitely often.[65] Does the argument work? How important is the role played in their argument by the many-worlds interpretation of quantum mechanics?

[63] See, e.g., Page, "Observational Selection Effects in Quantum Cosmology" and the references therein.
[64] See, e.g., Hollands and Wald, "Quantum Fields in Curved Spacetime," §2.2(b).
[65] Boddy, Carroll, and Pollack, "Why Boltzmann Brains Do Not Fluctuate into Existence from the de Sitter Vacuum." The argument is reprised—and some objections to it canvassed—in Carroll, "Why Boltzmann Brains are Bad," §4.

c) Inflationary cosmology offers an account of structure formation: the post-inflationary state was the Bunch-Davies vacuum; vacuum fluctuations provided the seeds for all of the non-uniform structure that we see today.[66] Are Boddy and co-authors in danger of proving too much—if their argument works against Boltzmann brains, will it also rule out the standard account of structure formation in the early Universe?

d) Gott leans heavily on the analogy between the Unruh effect and vacuum fluctuations in the static patch to argue that Boltzmann brains would not be physically real (although any photographs taken of them would be).[67] Reconstruct and evaluate his argument.

QUESTION 9.3 (Medium?). Boddy, Carrol, and Pollack make the alluring claim that:

The existence of Boltzmann Brain fluctuations is a rare example of a question whose answer depends sensitively on one's preferred formulation of quantum theory.[68]

Are they right?

QUESTION 9.4 (Larger). There is also a route to the problem of Boltzmann brains that does not pass through asymptotically de Sitter spacetimes. Consider the sort of universes one expects under the hypothesis of eternal inflation, consisting of infinitely many bubble universes with different initial conditions, continually coming into existence.[69] One expects that just about everything will happen somewhen and somewhere in such a universe. So it appears that we again have the fixings for the problem of Boltzmann Brains. Is that right? What are the advantages and disadvantages of this route to the problem, in contrast to the de Sitter route(s)?

QUESTION 9.5 (Larger!!). In a spatially compact universe that begins with a big bang and has an asymptotically de Sitter future, it is plausible that we

[66] For discussion and references, see Smeenk, "Inflation and the Origins of Structure."
[67] Gott, "Boltzmann Brains—I'd Rather See Than Be One."
[68] Boddy, Carroll, and Pollack, "Boltzmann Brains Do Not Fluctuate," 229.
[69] For overviews of the ideas behind eternal inflation, see Aguirre, "Cosmological Intimations of Infinity"; and Guth, "Eternal Inflation and Its Implications." For the difficulties of implementing these ideas within classical general relativity, see Ellis and Stoeger, "A Note on Infinities in Eternal Inflation." For further discussion, see Smeenk, "Testing Inflation."

can make comparisons like *genuine meetings of the Mathematical Association are far less common than mock meetings (of some specified type)*: let t be the cosmological time function (so t assigns to an event the number of units of proper time that have passed since the big bang according to fundamental observers); then calculate the number of events of each type that occur per unit of t, as $t \to \infty$. Can you find a similarly principled recipe for making such comparisons in under the hypothesis of eternal inflation?[70]

QUESTION 9.6 (Medium?). Is the following line of thought compelling?

If we put a kettle of water on the fire there is a chance that the water will freeze. If mankind goes on putting kettles on the fire until $t = \infty$, the chance will one day come off and the individual concerned will be somewhat surprised to find a lump of ice in his kettle. But it will not happen to *me*. Even if tomorrow the phenomenon occurs before my eyes I shall not explain it this way; I shall probably say the devil is in it. I would much sooner believe in the devil than in a coincidence of that kind coming off. And in doing so I shall be acting as a rational scientist. The reason why I do not at present believe that devils interfere with my cooking arrangements and other business, is because by observation I have become convinced by experience that Nature obeys certain uniformities which we call laws. I am convinced because these laws have been tested over and over again. But it is possible that every single observation from the beginning of science, which has been used as a test, has just happened to fit in with the law by a chance coincidence. It is an improbable coincidence, but I think not quite so improbable as the coincidence involved in my kettle of water freezing. So if the event happens and I can think of no other explanation, I shall have to choose between two highly improbable coincidences: (a) that there are no laws of Nature and that the apparent uniformities so far observed are merely coincidences, (b) that the event is entirely in accordance with the accepted laws of Nature, but that an improbable coincidence has happened. I choose the former because mathematical calculation indicates that it is the less improbable. You see that I reckon a sufficiently improbable coincidence as something much more disastrous than a violation of the laws of Nature;

[70] That is: can you solve the measure problem for eternal inflation? For a recent survey of extant approaches, see Winitzki, *Eternal Inflation*, chapter 6. For further discussion, see Smeenk, "Testing Inflation"; and Dorr and Arntzenius, "Self-Locating Priors and Cosmological Measures."

because my whole reason for accepting the laws of Nature rests on the assumption that improbable coincidences do not happen—at least that they do not happen in my experience.[71]

QUESTION 9.7 (Medium?). Tumulka works with a Bohmian quantum field theory in a cosmological patch of a de Sitter spacetime and reaches the conclusion that you should expect few if any copies of your current self to fluctuate out of the vacuum in the infinite future.[72] Should we expect this result to hold more generally for (well-behaved) quantum fields in the cosmological patch?

QUESTION 9.8 (Medium?). Suppose that we follow Dyson, Kleban, and Susskind in thinking that a cosmological model should be centred on a particular observer and should not include regions of spacetime that are out of causal contact with that observer. Dyson, Kleban, and Susskind limit their model to the static patch of the observer. Parikh, Savonije, Verlinde, who adopt a similar starting point, instead work with the cosmological patch of the observer (see Section 5.4 above). Which approach is better motivated?

QUESTION 9.9 (Larger). Suppose that theory T_b is afflicted with the problem of Boltzmann brains while theory T_a is not—but that the theories otherwise account equally well for our evidence and are otherwise comparably virtuous.

a) Many physicists and philosophers think that through ordinary experience we constantly acquire evidence that favours T_a over T_b: after all, it is a feature of T_b that typical observers whose situation was indiscernible from yours five minutes ago were in for some very nasty surprises—whereas things presumably went comparably well for you in the last five minutes. Is this common line of thought cogent? Does it depend on you having reliable memories of the last five minutes?

b) Carroll argues that there is a sense in which T_b is *cognitively unstable*: if we come to believe T_b on the basis of empirical evidence then our belief

[71] Eddington, "The End of the World," 321.
[72] Tumulka, "Long-Time Asymptotics of a Bohmian Scalar Quantum Field in de Sitter Space-Time."

in T_b should undermine our belief in this evidence.[73] He concludes that we should assign T_b prior to probability zero. Is that reasonable?

c) There are epistemological and ethical theories that are in a sense self-undermining.[74] How is Carroll's notion of cognitive instability related to such self-undermining? Should scientific and philosophical theories be held to the same standards in this regard?

QUESTION 9.10 (Medium?). In the physics literature, there is a great deal of focus on shallow fluctuations containing duplicates of human brains. It is usually asserted that such short-lived Boltzmann brains should be expected to have very strange and incoherent experiences and memories. Are there psychological bounds on how incoherent experiences and memories can be?

QUESTION 9.11 (Smaller?). Dogramaci offers the following sort of response to the problem of Boltzmann brains.[75] Our total evidence—the basis of our belief in the cosmological theories that are supposed to be the root of the problem—"includes in it lots that says we're in ordinary human bodies, on ordinary earth, which has existed and circled the sun for billions of years, and so on and so on." On the basis of this, it is reasonable to believe that we are not living in a shallow fluctuation from equilibrium. Of course, our near-duplicates who *do* live in shallow fluctuations have the same evidence—but in their case it is misleading. Rationality dictates that all of us should believe that we are ordinary observers rather than Boltzmann brains (BBs):

> We should all conclude, me and the BBs, that we are in the good case. All of the zillions of BBs should think they are not BBs. I will be right, they will be wrong, and we'll all be rational.

Does it follow, on Dogramaci's account, that whether our ordinary inductive procedures are reliable depends on whether or not $\Lambda > 0$?

QUESTION 9.12 (Smaller?). David Lewis crystallizes one popular response to skeptical arguments when he says that we know a lot—and, further, that:

---

[73] See Carroll, "Why Boltzmann Brains are Bad." For critical discussion, see Kotzen, "What Follows from the Possibility of Boltzmann Brains?"; and Wallace, "A Bayesian Analysis of Self-Undermining Arguments in Physics."

[74] For discussion and examples, see Railton, "Alienation, Consequentialism, and the Demands of Morality"; and Railton, "Truth, Reason, and the Regulation of Belief."

[75] Dogramaci, "Does My Total Evidence Support that I'm a Boltzmann Brain?"

It is a Moorean fact that we know a lot. It is one of those things that we know better than we know the premises of any philosophical argument to the contrary.[76]

Do those who take the Dyson-Kleban-Susskind route to the problem of Boltzmann brains need to maintain that we know the principle of observer complementarity better than we know that we know a great deal about the distant past?

QUESTION 9.13 (Larger!!!). What do you think about step (v) in the argument of Section 1?[77]

---

[76] Lewis, "Elusive Knowledge," 549. Moore takes a similar line in "Four Forms of Scepticism," 225 f.

[77] There is already a vast literature on this question. A few references to get started: Srednicki and Hartle, "Science in a Very Large Universe"; Srednicki and Hartle, "The Xerographic Distribution"; Bostrom, *Anthropic Bias*; Dorr and Arntzenius, "Self-Locating Priors"; and Manley, "On Being a Random Sample."

References

The Collected Papers of Albert Einstein. For the most part, Einstein's writings and correspondence are cited here in the version published in the ongoing Princeton University Press edition of *The Collected Papers of Albert Einstein*—and some of the editorial apparatus from that edition is also cited (this apparatus can be found only in the original-language volumes, not in the volumes consisting of English translations). Material by Einstein and his correspondents is cited by (English) title or date, followed by volume number and item number. Editorial apparatus is cited by volume number and further identifying details.

Abraham, R. and J. Marsden. *Foundations of Mechanics* (second edition). Benjamin-Cummings Publishing (1978).

Aguirre, A. "Cosmological Intimations of Infinity." In M. Heller and H. Woodin (eds.), *Infinity: New Research Frontiers.* Cambridge University Press (2011) 176–192.

Aguirre, A. and S. Gratton, "Inflation without a Beginning: A Null Boundary Proposal." *Physical Review D* 67 (2003) 083515.

Albert, D. *After Physics.* Harvard University Press (2015).

Albert, D. *Time and Chance.* Harvard University Press (2000).

Albrecht, A. "Tuning, Ergodicity, Equilibrium, and Cosmology." *Physical Review D* 91 (2015) 103510.

Albrecht, A. and L. Sorbo. "Can the Universe Afford Inflation?" *Physical Review D* 70 (2004) 063528.

Almheiri, A., N. Engelhardt, D. Marolf, and H. Maxfield. "The Entropy Bulk of Quantum Fields and the Entanglement Wedge of an Evaporating Black Hole." *Journal of High Energy Physics* 2019 (2019) 063.

Almheiri, A., D. Marolf, J. Polchinski, and J. Sully. "Black Holes: Complementarity or Firewalls?" *Journal of High Energy Physics* 2013 (2013) 062.

Anderson, M. "Existence and Stability of Even-Dimensional Asymptotically de Sitter Spaces." *Annales Henri Poincaré* 6 (2005) 801–820.

Anderson, M. "On Long-Time Evolution in General Relativity and Geometrization of 3-Manifolds." *Communications in Mathematical Physics* 222 (2001) 533–567.

Anderson, M. "On Stationary Vacuum Solutions to the Einstein Equations." *Annales Henri Poincaré* 1 (2000) 977–994.

Anderson, M. "On the Structure of Asymptotically de Sitter and Anti-de Sitter Spaces." *Advances in Theoretical and Mathematical Physics* 8 (2005) 861–893.

Anderson, M. "On the Uniqueness and Global Dynamics of AdS Spacetimes." *Classical and Quantum Gravity* 23 (2006) 6935–6953.

Anderson, M., P. Chruściel, and E. Delay. "Non-Trivial, Static, Geodesically Complete, Vacuum Space-Times with a Negative Cosmological Constant." *Journal of High Energy Physics* 2002 (2002) 063.

Anderson, M., P. Chruściel, and E. Delay. "Non-Trivial, Static, Geodesically Complete, Vacuum Space-Times with a Negative Cosmological Constant. II. $n \geq 5$." In O. Biquard (ed.), *The AdS/CFT Correspondence: Einstein Metrics and Their Conformal Boundaries*. European Mathematical Society (2005) 165–204.

Andersson, L. "Momenta and Reduction for General Relativity." *Journal of Geometry and Physics* 4 (1987) 289–314.

Andersson, L., T. Barbot, R. Benedetti, F. Bonsante, W. Goodman, F. Labourie, K. Scannell, and J.-M. Schlenker. "Notes on: 'Lorentz Spaces of Constant Curvature'." *Geometriae Dedicata* 126 (2007) 47–70.

Andréasson, H., D. Fajman, and M. Thaller. "Static Solutions to the Einstein-Vlasov System with a Non-Vanishing Cosmological Constant." *SIAM Journal of Mathematical Analysis* 47 (2015) 2657–2688.

Andréasson, H. and H. Ringström, "Proof of the Cosmic No-Hair Conjecture in the $\mathbb{T}^3$-Gowdy Symmetric Einstein-Vlasov Setting." *Journal of the European Mathematical Society* 18 (2016) 1565–1650.

Aneesh, P., S. Jahanur Hoque, and A. Virmani. "Conserved Charges in Asymptotically de Sitter Spacetimes." *Classical and Quantum Gravity* 36 (2019) 205008.

Arnold, V. I. *Mathematical Understanding of Nature: Essays on Amazing Physical Phenomena and Their Understanding by Mathematicians*. American Mathematical Society (2014).

Arnold, V. I. and B. Khesin. *Topological Methods in Hydrodynamics*. Springer (1998).

Ashtekar, A. *Asymptotic Quantization*. Bibliopolis (1987).

Ashtekar, A. "Geometry and Physics of Null Infinity." In L. Bieri and S.-T. Yau (eds.), *One Hundred Years of General Relativity: A Jubilee Volume on General Relativity and Mathematics*. International Press (2015) 99–122.

Ashtekar, A. "Implications of a Positive Cosmological Constant for General Relativity." *Reports on Progress in Physics* 80 (2017) 102901.

Ashtekar, A., and S. Bahrami. "Asymptotics with a Positive Cosmological Constant: IV. The No-Incoming Radiation Condition." *Physical Review D* 92 (2015) 044011.

Ashtekar, A., L. Bombelli, and O. Reula, "The Covariant Phase Space of Asymptotically Flat Gravitational Fields." In M. Francaviglia (ed.), *Mechanics, Analysis and Geometry: 200 Years after Lagrange*. North-Holland (1991) 417–450.

Ashtekar, A., B. Bonga, and A. Kesavan. "Asymptotics with a Positive Cosmological Constant: I. Basic Framework." *Classical Quantum Gravity* 32 (2015) 025004.

Ashtekar, A., B. Bonga, and A. Kesavan. "Asymptotics with a Positive Cosmological Constant: II. Linear Fields on de Sitter Spacetime." *Physical Review D* 92 (2015) 044011.

Ashtekar, A., B. Bonga, and A. Kesavan. "Asymptotics with a Positive Cosmological Constant: III. The Quadrapole Formula." *Physical Review D* 92 (2015) 104032.

Ashtekar, A., M. Campiglia, and A. Laddha. "Null Infinity, the BMS Group, and Infrared Issues." *General Relativity and Gravitation* 50 (2018) 140.

Ashtekar, A. and S. Das."Asymptotic Anti-de Sitter Spacetimes: Conserved Quantities." *Classical and Quantum Gravity* 17 (2000) L17–L30.

Ashtekar, A. and A. Magnon."Asymptotically Anti-de Sitter Space-Times." *Classical and Quantum Gravity* 1 (1984) L39–L45.

Bain, J. "The RT Formula and Its Discontents: Spacetime and Entanglement." *Synthese* 198 (2021) 11833–11860.

Banks, T. and L. Mannelli. "De Sitter Vacua, Renormalization, and Locality." *Physical Review D* 67 (2003) 065009.

Barrow, J., G. Ellis, R. Maartens, and C. Tsagas. "On the Stability of the Einstein Static Universe." *Classical and Quantum Gravity* 20 (2003) L155–L164.

Barrow, J. and F. Tipler. *The Anthropic Cosmological Principle*. Oxford University Press (1986).

Bartnik, R. and J. Isenberg. "The Constraint Equations." In P. Chruściel and H. Friedrich (eds.), *The Einstein Equations and the Large Scale Behavior of Gravitational Fields: 50 Years of the Cauchy Problem in General Relativity*. Birkhäuser (2004) 1–38.

Batterman, R. *A Middle Way: A Non-Fundamental Approach to Many-Body Physics*. Oxford University Press (2021).

Batterman, R. *The Devil in the Details: Asymptotic Reasoning in Explanation, Reduction, and Emergence*. Oxford University Press (2002).

Beem, J. "Homothetic Maps of the Space-Time Distance Function and Differentiability." *General Relativity and Gravitation* 9 (1978) 793–799.

Beem, J. "Lorentzian Distance and Curvature." In G. Rassias (ed.) *The Mathematical Heritage of C. F. Gauss*. World Scientific (1991) 53–65.

Beem, J., P. Ehrlich, and K. Easley. *Global Lorentzian Geometry* (second edition). CRC Press (1996).

Beig, R. and J. Heinzle. "CMC-Slicings of Kottler-Schwarzschild-de Sitter Cosmologies." *Communications in Mathematical Physics* 260 (2005) 673–709.

Beisbart, C. and T. Jung. "Privileged, Typical, or Not Even That?—Our Place in the World According to the Copernican and the Cosmological Principles." *Journal for General Philosophy of Science* 37 (2006) 225–256.

Belot, G. *Cosmological Reconsiderations*. Forthcoming. Some day.

Belot, G. "Dust, Time, and Symmetry." *The British Journal for the Philosophy of Science* 56 (2005) 255–291.

Belot, G. *Geometric Possibility*. Oxford University Press (2011).

Belot, G. "The Representation of Time and Change in Mechanics." In J. Earman and J. Butterfield (eds.), *Philosophy of Physics* Volume A. Elsevier (2007) 133–227.

Belot, G. "Transcendental Idealism among the Jersey Metaphysicians." *Philosophical Studies* 150 (2010) 429–438.

Belot, G., J. Earman, and L. Ruetsche. "The Hawking Information Loss Paradox: The Anatomy of a Controversy." *The British Journal for the Philosophy of Science* 50 (1999) 189–229.

Benedetti, R. and C. Petronio. *Lectures on Hyperbolic Geometry*. Springer (1992).

Berger, M. *A Panoramic View of Riemannian Geometry*. Springer (2003).

Bernal, A. and M. Sánchez. "Globally Hyperbolic Spacetimes Can Be Defined as 'Causal' Instead of 'Strongly Causal'." *Classical and Quantum Gravity* 24 (2007) 745–749.

Beyer, F. "Non-Genericity of the Nariai Solutions. I. Asymptotics and Spatially Homogeneous Perturbations." *Classical and Quantum Gravity* 26 (2009) 235015.

Beyer, F. "Non-Genericity of the Nariai Solutions. II. Investigations within the Gowdy Class." *Classical and Quantum Gravity* 26 (2009) 235016.

Bieri, L. "An Extension of the Stability Theorem of the Minkowski Space in General Relativity." *Journal of Differential Geometry* 86 (2010) 17–70.

Binney, J. and S. Tremaine. *Galactic Dynamics* (second edition). Princeton University Press (2008).

Birkhoff, G. "Metric Foundations of Geometry." *Transactions of the American Mathematical Society* 55 (1944) 465–492.

Bishop, R. and S. Goldberg. *Tensor Analysis on Manifolds*. Dover (1980).

Bizoń, P. and A. Rostworowski. "Weakly Turbulent Instability of Anti-de Sitter Spacetime." *Physical Review Letters* 107 (2011) 031102.

Blumenthal, L. "Congruence and Superposability in Elliptic Space." *Transactions of the American Mathematical Society* 62 (1947) 431–451.

Bocchieri, P. and A. Loinger. "Quantum Recurrence Theorem." *Physical Review* 107 (1957) 337–338.

Boddy, K., S. Carroll, and J. Pollack. "Why Boltzmann Brains Do Not Fluctuate into Existence from the de Sitter Vacuum." In K. Chamcham, J. Silk, J. Barrow, and S. Saunders (eds.), *The Philosophy of Cosmology*. Cambridge University Press (2017) 228–240.

Boltzmann, L. "On Certain Questions of the Theory of Gases." *Nature* 51 (1895) 413–415.

Boltzmann, L. "On Zermelo's Paper 'On the Mechanical Explanation of Irreversible Processes'." In S. Brush (ed.), *Kinetic Theory* Volume 2. Oxford University Press (1966) 238–245. Originally published as "Zu Hrn. Zermelo's Abhandlung ,Uber die mechanische Erklärung irreversibler Vorgange'." *Annalen der Physik* 60 (1897) 392–398.

Bondi, H. *Cosmology*. Cambridge University Press (1952).

Bonsante, F. and A. Seppi. "Anti-de Sitter Geometry and Teichmüller Theory." In V. Alberge, K. Ohshika, and A. Papadopoulos (eds.), *In the Tradition of Thurston: Geometry and Topology*. Springer (2020) 545–643.

Bostrom, N. *Anthropic Bias: Observation Selection Effects in Science and Philosophy*. Routledge (2002).

Brody, D. and L. Hughston. "Geometric Quantum Mechanics." *Journal of Geometry and Physics* 38 (2001) 19–53.

Brody, D. and L. Hughston. "Unitarity, Ergodicity, and Quantum Thermodynamics." *Journal of Physics A* 40 (2007) F503–F509.

Calabi, E. and L. Markus. "Relativistic Space Forms." *Annals of Mathematics* 75 (1962) 63–76.

Callahan, J. *The Geometry of Spacetime: An Introduction to Special and General Relativity*. Springer (2000).

Carrera, M. and D. Giulini. "Influence of Global Cosmological Expansion in Local Dynamics and Kinematics." *Reviews of Modern Physics* 82 (2010) 169–208.

Carroll, S. "In What Sense Is the Early Universe Fine-Tuned?" In B. Loewer, E. Winsberg, and B. Weslake (eds.), *The Probability Map of the Universe: Essays on David Albert's* Time and Chance. Harvard University Press (2023) 110–141.

Carroll, S. *Spacetime and Geometry: An Introduction to General Relativity.* Addison Wesley (2004).

Carroll, S. "Why Boltzmann Brains are Bad." In S. Dasgupta, R. Dotan, and B. Weslake (eds.), *Current Controversies in Philosophy of Science.* Routledge (2021) 7–20.

Cercignani, C., R. Illner, and M. Pulvirenti. *The Mathematical Theory of Dilute Gases.* Springer (1994).

Chen, B.-L. "On Stationary Solutions to the Vacuum Einstein Field Equations." *Asian Journal of Mathematics* 23 (2019) 609–630.

Choquet-Bruhat, Y. *A Lady Mathematician in this Strange Universe: Memoirs.* World Scientific (2018).

Choquet-Bruhat, Y. *General Relativity and the Einstein Equations.* Oxford University Press (2009).

Choquet-Bruhat, Y. *Introduction to General Relativity, Black Holes, and Cosmology.* Oxford University Press (2015).

Christodoulou, D. *Mathematical Problems of General Relativity* Volume I. European Mathematical Society (2008).

Christodoulou, D. "On the Global Initial Value Problem and the Issue of Singularities." *Classical and Quantum Gravity* 16 (1999) A23–A35.

Christodoulou, D. and S. Klainerman. *The Global Nonlinear Stability of the Minkowski Space.* Princeton University Press (1993).

Cinti, E. and V. Fano. "Careful with Those Scissors, Eugene! Against the Observational Indistinguishability of Spacetimes." *Studies in History and Philosophy of Science* 89 (2021) 103–113.

Cohen, B. and A. Whitman (eds.) *Isaac Newton: The Principia.* University of California Press (1999).

Compère, G., A. Fiorucci, and R. Ruzziconi. "The Λ-BMS$_4$ Group of dS$_4$ and New Boundary Conditions for AdS$_4$." *Classical and Quantum Gravity* 36 (2019) 195017.

Conway, J., H. Burgiel, and C. Goodman-Strauss. *The Symmetries of Things.* CRC Press (2008).

Coudène, Y. *Ergodic Theory and Dynamical Systems.* Springer (2016).

Coxeter, H. S. M. "A Geometrical Background for De Sitter's World." *The American Mathematical Monthly* 50 (1943) 217–228.

Coxeter, H. S. M. *Non-Euclidean Geometry* (fifth edition). University of Toronto Press (1965).

Crampin, M. and F. Pirani. *Applicable Differential Geometry.* Cambridge University Press (1986).

Cushman, R. and L. Bates. *Global Aspects of Classical Integrable Systems.* Birkhäuser (1997).

Dafermos, C. *Hyperbolic Conservation Laws in Continuum Physics* (fourth edition). Springer (2016).

Dafermos, M. "The Formation of Black Holes in General Relativity (After D. Christodoulou)." *Astérisque* 352 (2013) 243–313.

Dafermos, M. and G. Holzegel. "Dynamic Instability of Solitons in $4+1$-Dimensional Gravity with Negative Cosmological Constant." Unpublished manuscript (2006).

D'Agostino, S. "The *Bild* Conception of Physical Theory: Helmholtz, Hertz, and Schrödinger." *Physics in Perspective* 6 (2004) 372–389.

Darrigol, O. *Atoms, Mechanics, and Probability: Ludwig Boltzmann's Statistico-Mechanical Writings.* Oxford University Press (2018).

Davidson, D. "Knowing One's Own Mind." *Proceedings and Addresses of the American Philosophical Association* 60 (1987) 441–458.

de Rham, G. "Complexes à automorphismes et homéomorphie différentiable." *Annales de l'Institut Fourier* 2 (1950) 51–67.

de Sitter, W. "Further Remarks on the Solutions of the Field-Equations of Einstein's Theory of Gravitation." *Koninklijke Akademie van Wetenschappen te Amsterdam. Section of Sciences. Proceedings* XX (1918) 1309–1312. Originally published as "Nadere opmerkingen de oplossingen der veldvergelijkingen van Einstein's gravitatie-theorie." *Koninklijke Akademie van Wetenschappen te Amsterdam. Wis-en Natuurkundige Afdeeling. Verslagen van de Gewone Veraderingen* 26 (1917–1918) 1472–1475.

de Sitter, W. "On the Relativity of Inertia: Remarks Concerning Einstein's Latest Hypothesis." *Koninklijke Akademie van Wetenschappen te Amsterdam: Section of Sciences. Proceedings* XIX (1917) 1217–1225. Originally published as "Over de relativiteit der traagheid. Beschouwingen naar aanleiding van Einstein's laatste hypothese." *Koninklijke Akademie van Wetenschappen te Amsterdam. Wis-en Natuurkundige Afdeeling. Verslagen van de Gewone Veraderingen* 25 (1917) 1217–1225.

Dieks, D. "Communication by EPR devices." *Physics Letters A* 92 (1982) 271–272.

Dirac, P. A. M. "The Electron Wave Equation in de-Sitter Space." *Annals of Mathematics* 36 (1935) 657–669.

Di Valentina, E., O. Mena, S. Pan, L. Visinelli, W. Yang, A. Melchiorri, D. Mota, A. Reiss, and J. Silk. "In the Realm of the Hubble Tension—a Review of Solutions." *Classical and Quantum Gravity* 38 (2021) 153001.

Doboszewski, J. "Interpreting Cosmic No Hair Theorems: Is Fatalism about the Far Future of Expanding Cosmological Models Unavoidable?" *Studies in History and Philosophy of Modern Physics* 66 (2019) 170–179.

Dodelson, S. and F. Schmidt. *Modern Cosmology* (second edition). Elsevier (2020).

Dogramaci, S. "Does My Total Evidence Support that I'm a Boltzmann Brain?" *Philosophical Studies* 177 (2020) 3717–3723.

Dorr, C. and F. Arntzenius. "Self-Locating Priors and Cosmological Measures." In K. Chamcham, J. Silk, J. Barrow, and S. Saunders (eds.), *The Philosophy of Cosmology.* Cambridge University Press (2017) 396–428.

Dumas, H. *The KAM Story: A Friendly Introduction to the Content, History, and Significance of Classical Kolmogorov-Arnold-Moser Theory.* World Scientific (2014).

Duncan, D. and E. Ihrig. "Homogeneous Spaces of Zero Curvature." *Proceedings of the American Mathematical Society* 107 (1989) 785–795.

Du Val, P. "Geometrical Note on de Sitter's World." *Philosophical Magazine* 47 (1924) 930–938.

Du Val, P. "On the Discriminations between Past and Future." *Philosophical Magazine* 49 (1925) 379–390.

Dyson, L. *Three Lessons in Causality: What String Theory Has to Say About Naked Singularities, Time Travel, and Horizon Complementarity.* Unpublished MIT doctoral dissertation (2004).

Dyson, L., M. Kleban, and L. Susskind. "Disturbing Implications of a Cosmological Constant." *Journal of High Energy Physics* 2002 (2002) 011.

Eardley, D., J. Isenberg, J. Marsden, and V. Moncrief. "Homothetic and Conformal Symmetries of Solutions to Einstein's Equations." *Communications in Mathematical Physics* 106 (1986) 137–158.

Earman, J. *Bangs, Crunches, Whimpers, and Shrieks: Singularities and Acausalities in Relativistic Spacetimes.* Oxford University Press (1995).

Earman, J. "Underdetermination, Realism, and Reason." *Midwest Studies in Philosophy* XVIII (1993) 19–38.

Earman, J. "What Time Reversal Is and Why It Matters." *International Studies in Philosophy of Science* 16 (2002) 245–264.

Earman, J. and J. Norton, "Forever Is a Day: Supertasks in Pitowsky and Malament-Hogarth Spacetimes." *Philosophy of Science* 60 (1993) 22–42.

Eddington, A. *New Pathways in Science.* MacMillan and Company (1935).

Eddington, A. "On the Instability of Einstein's Spherical World." *Monthly Notices of the Royal Astronomical Society* 90 (1930) 668–678.

Eddington, A. *Space, Time, and Gravitation.* Cambridge University Press (1921).

Eddington, A. "The End of the World: From the Standpoint of Mathematical Physics." *The Mathematical Gazette* 15 (1931) 316–324.

Egorov, Y. and M. Shubin. *Foundations of the Classical Theory of Partial Differential Equations.* Springer (1998).

Eichhorn, J. *Global Analysis on Open Manifolds.* Nova (2007).

Einstein, A. "On the So-Called Cosmological Problem." Published as an appendix to C. O'Raifeartaigh, M. O'Keeffe, W. Nahm, and S. Mitton, "Einstein's Cosmology Review of 1933: A New Perspective on the Einstein-de Sitter Model of the Cosmos." *The European Physics Journal H* 40 (2015) 301–335. Originally published as "Sur la structure cosmologique de l'espace." In M. Solovine (ed.), *Les fondements de la théorie de la relativité générale. Théorie unitatire de la gravitation et de l'electricité. Sur la structure cosmologique de l'espace.* Hermann (1933) 99–109.

Einstein, A. "Zum kosmologischen Problem der allgemeinen Relativitätstheorie." *Sitzungsberichte der Preussischen Akademie der Wissenschaften* 1931 (1931) 235–237.

Eisenhart, L. *Riemannian Geometry.* Princeton University Press (1925).

Ellis, G. "Topology and Cosmology." *General Relativity and Gravitation* 2 (1971) 7–21.

Ellis, G. and W. Stoeger. "A Note on Infinities in Eternal Inflation." *General Relativity and Gravitation* 41 (2009) 1475–1484.

Epple, M. "From Quaternions to Cosmology: Spaces of Constant Curvature, ca. 1873–1925." In T. Li (ed.), *Proceedings of the International Congress of Mathematicians* Volume III. Higher Education Press (2002) 935–945.

Eshkobilov, O., E. Musso, and L. Nicolodi. "On the Restricted Conformal Group of the $(1 + n)$-Einstein Static Universe." *Journal of Geometry and Physics* 146 (2019) 103517.

Evans, L. *Partial Differential Equations* (second edition). American Mathematical Society (2010).

Feynman, R. *The Character of Physical Law*. MIT Press (1967).

Feynman, R., R. Leighton, and M. Sands. *The Feynman Lectures on Physics*. Addison-Wesley (1963).

Fichant, M. (ed.) *De l'horizon de la doctrine humaine (1693). Ἀποκατάστασις πάντων (La restitution universelle) (1715)*. Vrin (1991).

Fine, K. "Tense and Reality." In K. Fine, *Modality and Tense: Philosophical Papers*. Oxford University Press (2005) 261–320.

Fischer, A. and V. Moncrief. "The Reduced Einstein Equations and the Conformal Volume Collapse of 3-Manifolds." *Classical and Quantum Gravity* 18 (2001) 4493–4515.

Fletcher, S. "Similarity, Topology, and Physical Significance in Relativity Theory." *The British Journal for the Philosophy of Science* 67 (2016) 365–389.

Fletcher, S. "The Principle of Stability." *Philosophers' Imprint* 20 (2020) 03.

Fletcher, S., J. Manchak, M. Schneider, and J. Weatherall. "Would Two Dimensions Be World Enough for Spacetime?" *Studies in History and Philosophy of Modern Physics* 63 (2018) 100–113.

Flin, P. and H. Duerbeck. "Silberstein, Relativity, and Cosmology." In H.-M. Alimi and A. Füzfa (eds.), *Albert Einstein Century International Conference*. American Institute of Physics (2006) 1087–1094.

Forman, D. "Leibniz on Human Finitude, Progress, and Eternal Recurrence: The Argument of the 'Apokatastasis' Essay Drafts and Related Texts." *Oxford Studies in Early Modern Philosophy* VIII (2018) 225–270.

Frances, C. "The Conformal Boundary of Anti-de Sitter Spacetimes." In O. Biquard (ed.), *The AdS/CFT Correspondence: Einstein Metrics and Their Conformal Boundaries*. European Mathematical Society (2005) 205–216.

Frauendiener, J. "Conformal Infinity—Development and Applications." In D. Rowe, T. Sauer, and S. Walter (eds.), *Beyond Einstein: Perspectives on Geometry, Gravitation, and Cosmology in the Twentieth Century*. Birkhäuser (2018) 451–473.

Friedman, J. and A. Higuchi. "Quantum Field Theory in Lorentzian Universes from Nothing." *Physical Review D* 52 (1995) 5687–5697.

Friedmann, A. "On the Possibility of a World with Constant Negative Curvature of Space." *General Relativity and Gravitation* 31 (1999) 2001–2008. Originally published as "Uber die Möglichkeit einer Welt mit konstanter negativer Krümmung des Raumes." *Zeitschrift für Physik* 21 (1924) 326–332.

Friedrich, H. "Einstein Equations and Conformal Structure: Existence of Anti-de Sitter-Type Space-Times." *Journal of Geometry and Physics* 17 (1995) 125–184.

Friedrich, H. "Geometric Asymptotics and Beyond." In L. Bieri and S.-T. Yau (eds.), *One Hundred Years of General Relativity: A Jubilee Volume on General Relativity and Mathematics*. International Press (2015) 37–74.

Friedrich, H. "On the AdS Stability Problem." *Classical and Quantum Gravity* 31 (2014) 105001.

Frigg, R. and C. Werndl. "Explaining Thermodynamic-Like Behavior in Terms of Epsilon-Ergodicity." *Philosophy of Science* 78 (2011) 628–652.

Gallavotti, G. *Statistical Mechanics: A Short Treatise.* Springer (1999).

Gallot, S., D. Hulin, and J. Lafontaine. *Riemannian Geometry* (third edition). Springer (2004).

Galloway, G. "Some Global Results for Asymptotically Simple Space-Times." In J. Frauendiener and H. Friedrich (eds.), *The Conformal Structure of Space-Time: Geometry, Analysis, Numerics.* Springer (2002) 51–60.

Gao, S. and R. Wald. "Theorems of Gravitational Time Delay and Related Issues." *Classical and Quantum Gravity* 17 (2000) 4999–5008.

Garfinkle, D. and Q. Tian. "Spacetimes with Cosmological Constant and a Conformal Killing Field Have Constant Curvature." *Classical and Quantum Gravity* 4 (1987) 137–139.

Geroch, R. "Asymptotic Structure of Space-Time." In P. Esposito and L. Witten (eds.), *Asymptotic Structure of Space-Time.* Plenum (1977) 1–105.

Geroch, R. "General Relativity in the Large." *General Relativity and Gravitation* 2 (1971) 61–74.

Gibbons, G. "The Elliptic Interpretation of Black Holes and Quantum Mechanics." *Nuclear Physics B* 271 (1986) 497–508.

Gibbons, G. and S. Hawking. "Cosmological Event Horizons, Thermodynamics, and Particle Creation." *Physical Review D* 15 (1977) 2738–2751.

Gibbs, J. *Elementary Principles in Statistical Mechanics: Developed with Especial Reference to the Rational Foundations of Thermodynamics.* Charles Scribner's Sons (1902).

Girbau, J. and L. Bruna. *Stability by Linearization of Einstein's Field Equation.* Birkhäuser (2010).

Giulini, D. "Uniqueness of Simultaneity." *The British Journal for the Philosophy of Science* 52 (2001) 651–670.

Glymour, C. "Indistinguishable Space-Times and the Fundamental Group." In J. Earman, C. Glymour, and J. Stachel (eds.), *Foundations of Space-Time Theories.* University of Minnesota Press (1977) 50–60.

Glymour, C. *Theory and Evidence.* Princeton University Press (1980).

Glymour, C. "Topology, Cosmology, and Convention." *Synthese* 24 (1972) 195–218.

Gödel, K. "A Remark about the Relationship between Relativity Theory and Idealistic Philosophy." In S. Feferman, J. Dawson. S. Kleene, G. Moore, R. Solovay, and J. can Heijenhoort (eds.), *Kurt Gödel: Collected Works* Volume II. Oxford University Press (1990) 202–207. This is a somewhat expanded version of the original in P. Schilpp (ed.), *Albert Einstein: Philosopher-Scientist.* Open Court (1949) 557–562.

Goenner, H. "Weyl's Contributions to Cosmology." In E. Scholz (ed.), *Hermann Weyl's Raum-Zeit-Materie and a General Introduction to His Scientific Work.* Springer (2001) 105–137.

Goldstein, S. and J. Lebowitz. "On the (Boltzmann) Entropy of Non-Equilibrium Systems." *Physica D* 193 (2004) 53–66.

Goldstein, S. D. Huse, J. Lebowitz, and R. Tumulka. "Macroscopic and Microscopic Thermal Equilibrium." *Annalen der Physik* 529 (2017) 1600301.

Goldstein, S., J. Lebowitz, R. Tumulka, and N. Zanghì. "Long-Time Behavior of Macroscopic Quantum Systems: Commentary Accompanying the English

Translation of von Neumann's 1929 Article on the Quantum Ergodic Theorem." *The European Physical Journal H* 35 (2010) 173–200.

Gott, R. "Boltzmann Brains—I'd Rather See Than Be One." arXiv:0802.0233

Griffiths, J. and J. Podolský. *Exact Space-Times in Einstein's General Relativity.* Cambridge University Press (2009).

Gromov, M. *Metric Structures for Riemannian and Non-Riemannian Spaces.* Birkhäuser (2001).

Grünbaum, A. *Philosophical Problems of Space and Time* (second edition). Reidel (1973).

Guth, A. "Eternal Inflation and Its Implications." *Journal of Physics A* 40 (2007) 6811–6826.

Gyenis, B. *Well Posedness and Physical Possibility.* Unpublished University of Pittsburgh doctoral dissertation (2013).

Hacking, I. "The Identity of Indiscernibles." *Journal of Philosophy* LXXII (1975) 249–256.

Hackl, L. and Y. Neiman. "Horizon Complementarity in Elliptic de Sitter Space." *Physical Review D* 91 (2015) 044016.

Hadamard, J. "Les problèmes aux limites dans la théorie des équations aux dérivées partielles." *Journal de Physique Théorique et Appliquée* 6 (1907) 202–241.

Hall, G. *Symmetries and Curvature Structure in General Relativity.* World Scientific (2004).

Harlow, D. "Jerusalem Lectures on Black Holes and Quantum Information." *Reviews of Modern Physics* 88 (2016) 015002.

Hastie, T., R. Tibshirani, and J. Friedman. *The Elements of Statistical Learning: Data Mining, Inference, and Prediction* (second edition). Springer (2009).

Havas, P. "The General-Relativistic Two-Body Problem and the Einstein-Silberstein Controversy." In J. Earman, M. Janssen, and J. Norton (eds.), *The Attraction of Gravitation: New Studies in the History of General Relativity.* Birkhäuser (1993) 88–125.

Hawking, S. "Breakdown of Predictability in Gravitational Collapse." *Physical Review D* 14 (1976) 2460–2473.

Hawking, S. "Particle Creation by Black Holes." *Communications in Mathematical Physics* 43 (1975) 199–220.

Hawking, S. "Stable and Generic Properties in General Relativity." *General Relativity and Gravitation* 1 (1971) 393–400.

Hawking, S. "The Boundary Conditions for Gauged Supergravity." *Physics Letters* 126B (1983) 175–177.

Hawking, S. and G. Ellis (1973) *The Large Scale Structure of Space-Time.* Cambridge University Press.

Hawking, S., M. Perry, and A. Strominger. "Soft Hair on Black Holes." *Physical Review Letters* 116 (2016) 231301.

Hawking, S., M. Perry, and A. Strominger. "Superrotation Charge and Supertranslation Hair on Black Holes." *Journal of High Energy Physics* 2017 (2017) 161.

Hawkins, T. *Emergence of the Theory of Lie Groups: An Essay in the History of Mathematics 1869–1926.* Springer (2000).

Hayden, P. and J. Preskill. "Black Holes as Mirrors: Quantum Information in Random Subsystems." *Journal of High Energy Physics* 2007 (2007) 120.

Henneaux, M. and C. Teitelboim. "Asymptotically Anti-de Sitter Spaces." *Communications in Mathematical Physics* 98 (1985) 391–424.

Hilbert, D. "The Foundations of Physics (Second Communication)." In J. Renn (ed.), *The Genesis of General Relativity* Volume 4. Springer (2007) 1017–1038. Originally published as "Die Grundlagen der Physik (Zweite Mitteilung)." *Nachrichten von der Gesellschaft der Wissenschaften zu Göttingen. Mathematisch-Physikalische Klasse* 1917 (1917) 53–76.

Hilbert, D. and S. Cohn-Vossen. *Geometry and the Imagination.* Chelsea (1952).

Hinks, A. "West Vancouver Nonagenarian Sets Swimming World Records." *North Shore News* 29 January 2014.

Hintz, P. and A. Vasy. "The Global Non-Linear Stability of the Kerr-de Sitter Family of Black Holes." *Acta Mathematica* 220 (2018) 1–206.

Hirsch, M., S. Smale, and R. Devaney. *Differential Equations, Dynamical Systems, and an Introduction to Chaos* (third edition). Academic Press (2013).

Holfter, G. and H. Dickel. *An Irish Sanctuary: German-Speaking Refugees in Ireland 1933–1945.* Walter de Gruyter (2017).

Hollands, S. "Correlators, Feynman Diagrams, and Quantum No-Hair in de Sitter Spacetime." *Communications in Mathematical Physics* 319 (2013) 1–68.

Hollands, S. and R. Wald. "Quantum Fields in Curved Spacetime." *Physics Reports* 574 (2015) 1–35.

Holzegel, G., J. Luk, J. Smulevici, and C. Warnick. "Asymptotic Properties of Linear Field Equations in Anti-de Sitter Space." *Communications in Mathematical Physics* 374 (2020) 1125–1178.

Hounnonkpe, R. and E. Minguzzi. "Globally Hyperbolic Spacetimes Can Be Defined without the 'Causal' Condition." *Classical and Quantum Gravity* 36 (2019) 197001.

Hubeny, V. "The AdS/CFT Correspondence." *Classical and Quantum Gravity* 32 (2015) 124010.

Huggett, N. and C. Wüthrich. "Out of Nowhere: Duality." arXiv:2005.12728

Isakov, V. *Inverse Problems for Partial Differential Equations* (second edition). Springer (2006).

Ishibashi, A. and R. Wald. "Dynamics in Non-Globally-Hyperbolic Static Spacetimes: III. Anti-de Sitter Spacetime." *Classical and Quantum Gravity* 21 (2004) 2981–3013.

Janiak, A. (ed). *Newton: Philosophical Writings.* Cambridge University Press (2014).

Janssen, M. " 'No Success Like Failure…': Einstein's Quest for General Relativity, 1907–1920." In M. Janssen and C. Lehner (eds.), *The Cambridge Companion to Einstein.* Cambridge University Press (2014) 167–227.

Jeans, J. "Man and the Universe." In J. Jeans, W. Bragg, E. Appleton, E. Mellenby, J. Haldane, and J. Huxley, *Scientific Progress.* MacMillan (1936) 11–38.

John, F. *Partial Differential Equations* (fourth edition). Springer (1982).

Kastor, D. and J. Traschen, "A Positive Energy Theorem for Asymptotically de Sitter Spacetimes." *Classical and Quantum Gravity* 23 (2002) 5901–5920.

Kay, B. and R. Wald. "Theorems on the Uniqueness and Thermal Properties of Stationary, Nonsingular, Quasifree States on Spacetimes with a Bifurcate Killing Horizon." *Physics Reports* 207 (1991) 49–136.

Kehrberger, L. "The Case Against Smooth Null Infinity I: Heuristics and Counter-Examples." *Annales Henri Poincaré* 23 (2022) 829–921.

Kelly, W. and D. Marolf, "Phase Spaces for Asymptotically de Sitter Cosmologies." *Classical and Quantum Gravity* 29 (2012) 205013.

Kennefick, D. *Traveling at the Speed of Thought: Einstein and the Quest for Gravitational Waves.* Princeton University Press (2007).

Kiessling, K.-H. "The Influence of Gravity on the Boltzmann Entropy of a Closed Universe." In V. Allori (ed.), *Statistical Mechanics and Scientific Explanation: Determinism, Indeterminism and Laws of Nature.* World Scientific (2020) 387–419.

Kesavan, A. *Asymptotic Structure of Space-Time with a Positive Cosmological Constant.* Unpublished Pennsylvania State University doctoral dissertation (2016).

Keyfitz, B. "Shocks." In N. Higham, M. Dennis, P. Glendinning, P. Martin, F. Santosa, and J. Tanner (eds.), *The Princeton Companion to Applied Mathematics.* Princeton University Press (2015) 122–124.

Killing, W. *Die nicht-euklidischen Raumformen in analytischer Behandlung.* Teubner (1885).

Klainerman, S. and F. Nicolò. "Peeling Properties of Asymptotically Flat Solutions to the Einstein Vacuum Equations." *Classical and Quantum Gravity* 20 (2003) 3215–3257.

Klainerman, S. and F. Nicolò. *The Evolution Problem in General Relativity.* Birkhäuser (2003).

Klingler, B. "Complétude des variétés Lorentziennes à courbure constante." *Mathematische Annalen* 306 (1996) 353–370.

Kobayashi, S. *Transformation Groups in Differential Geometry.* Springer (1972).

Kotzen, M. "What Follows from the Possibility of Boltzmann Brains?" In S. Dasgupta, R. Dotan, and B. Weslake (eds.), *Current Controversies in Philosophy of Science.* Routledge (2021) 21–34.

Kragh, H. *Cosmology and Controversy: The Historical Development of Two Theories of the Universe.* Princeton University Press (1996).

Kunzinger, M. and C. Sämann. "Lorentzian Length Spaces." *Annals of Global Analysis and Geometry* 54 (2018) 399–447.

Lanczos, C. "On the Problem of Regular Solutions of Einstein's Gravitational Equations." In W. Davis (ed.), *Cornelius Lanczos: Collected Published Papers With Commentaries* Volume II. College of Physical and Mathematical Sciences North Carolina State University (1998) 734–765. Originally published as "Zur Frage der regulären Lösungen der Einsteinschen Gravitationsgleichungen." *Annalen der Physik* 13 (1932) 621–635.

Le Bihan, B. and J. Read. "Duality and Ontology." *Philosophy Compass* 13 (2018) e12555.

Lebowitz, J. "From Time-Symmetric Microscopic Dynamics to Time-Asymmetric Macroscopic Behavior: An Overview." In G. Gallavotti, W. Reiter, and J. Yngvason (eds.), *Boltzmann's Legacy.* European Mathematical Society (2008) 63–87.

Lee, J. *Introduction to Riemannian Manifolds* (second edition). Springer (2018).

Lee, J. *Manifolds and Differential Geometry.* American Mathematical Society (2009).

LeFloch, P. and J. Smulevici. "Future Asymptotics and Geodesic Completeness of Polarized T^2-Symmetric Spacetimes." *Analysis and PDE* 9 (2016) 363–395.

Lemaître, G. "A Homogeneous Universe of Constant Mass and Increasing Radius Accounting for the Radial Velocity of Extra-Galactic Nebulae." *General Relativity and Gravitation* 45 (2013) 1635–1646. Originally published as "Un univers homogène de masse constante et de rayon croissant, rendant compte de la vitesse radiale des nébuleuses extra-galactiques." *Annales de la Société Scientifique de Bruxelles* 47A (1927) 49–59.

Lesourd, M. "Observations and Predictions from Past Lightcones." *Classical and Quantum Gravity* 38 (2021) 115015.

Lewis, D. "Causation as Influence." *Journal of Philosophy* XCVII (2000) 182–197.

Lewis, D. "Elusive Knowledge." *Australasian Journal of Philosophy* 74 (1996) 549–567.

Lindblad, H. and I. Rodianski. "The Global Stability of Minkowski Space-Time in Harmonic Gauge." *Annals of Mathematics* 171 (2010) 1401–1477.

Louko, J. and D. Marolf, "Inextendible Schwarzschild Black Hole with a Single Exterior: How Thermal is the Hawking Radiation?" *Physical Review D* 58 (1998) 024007.

Luminet, J.-P. "Editorial Note to: Georges Lemaître, A Homogeneous Universe of Constant Mass and Increasing Radius Accounting for the Radial Velocity of Extra-Galactic Nebulae." *General Relativity and Gravitation* 45 (2013) 1619–1633.

Lynden-Bell, D. and R. Wood. "The Gravo-Thermal Catastrophe in Isothermal Spheres and the Onset of Red-Giant Structure for Stellar Systems." *Monthly Notices of the Royal Astronomical Society* 138 (1968) 495–525.

Maddy, P. *What Do Philosophers Do? Skepticism and the Practice of Philosophy.* Oxford University Press (2017).

Malament, D. "Observationally Indistinguishable Space-Times." In J. Earman, C. Glymour, and J. Stachel (eds.), *Foundations of Space-Time Theories.* University of Minnesota Press (1977) 61–80.

Maldacena, J. "Eternal Black Holes in Anti-de Sitter." *Journal of High Energy Physics* 2003 (2003) 021.

Maldacena, J. "The Large N Limit of Superconformal Field Theories and Supergravity." *International Journal of Theoretical Physics* 38 (1999) 1113–1133.

Manchak, J. B. "Can We Know the Global Structure of Spacetime?" *Studies in History and Philosophy of Modern Physics* 40 (2009) 53–56.

Manchak, J. B. *Global Spacetime Structure.* Cambridge University Press (2020).

Manchak, J. B. "What Is a Physically Reasonable Space-Time?" *Philosophy of Science* 78 (2011) 410–420.

Manley, D. "On Being a Random Sample." Unpublished.

Markus, L. and K. Meyer. *Generic Hamiltonian Dynamical Systems Are Neither Integrable Nor Ergodic.* American Mathematical Society (1974).

Marsden, J. and J. Isenberg. "A Slice Theorem for the Space of Solutions of Einstein's Equations." *Physics Reports* 89 (1982) 179–222.

Marsden, J. and T. Ratiu. "Nonlinear Stability in Fluids and Plasmas." In A. Tromba (ed.), *Seminar on New Results in Nonlinear Partial Differential Equations*. Friedrich Vieweg und Sohn (1987) 101–134.

Maudlin, T. *Philosophy of Physics: Space and Time*. Princeton University Press (2012).

McInnes, B. "De Sitter and Schwarzschild-de Sitter according to Schwarzschild and de Sitter." *Journal of High Energy Physics* 2003 (2003) 009.

Merleau-Ponty, J. *Cosmologie du XXᵉ siècle. Étude épistémologique et historique des théories de la cosmologie contemporaine*. Éditions Gallimard (1966).

Mess, G. "Lorentz Spaces of Constant Curvature." *Geometriae Dedicata* 126 (2007) 3–45.

Millikan, R. "On Knowing the Meaning: With a Coda on Swampman." *Mind* 119 (2010) 43–81.

Misner, C., K. Thorne, and J. Wheeler. *Gravitation*. W. H. Freeman and Company (1973).

Monclair, D. "Isometries of Lorentz Surfaces and Convergence Groups." *Mathematische Annalen* 363 (2015) 101–141.

Moncrief, V. and P. Mondal. "Could the Universe Have an Exotic Topology?" *Pure and Applied Mathematics Quarterly* 15 (2019) 921–966.

Moncrief, V. and P. Mondal. "Einstein Flow with Matter Sources: Stability and Convergence." *Philosophical Transactions of the Royal Society A* 380 (2022) 20210190.

Mondal, P. "The Nonlinear Stability of $n+1$-Dimensional FLRW Spacetimes." arXiv:2203.04785

Moore, C. "Ergodic Theorem, Ergodic Theory, and Statistical Mechanics." *Proceedings of the National Academy of Science* 112 (2015) 1907–1911.

Moore, G. E. "Four Forms of Scepticism." In G. E. Moore, *Philosophical Papers*. George Allen and Unwin (1959) 196–226.

Moore, W. *Schrödinger: Life and Thought*. Cambridge University Press (1989).

Morgan, J. "Recent Progress on the Poincaré Conjecture and the Classification of 3-Manifolds." *Bulletin of the American Mathematical Society* 42 (2004) 57–78.

Moschella, U. "The de Sitter and Anti-de Sitter Sightseeing Tour." *Séminaire Poincaré* 1 (2005) 1–12.

Moschidis, G. "A Proof of the Instability of AdS for the Einstein-Massless Vlasov System." *Inventiones Mathematicae* 231 (2023) 467–672.

Moschidis, G. "The Characteristic Initial-Boundary Value Problem for the Einstein-Massless Vlasov System in Spherical Symmetry." arXiv:1812.04274

Moschidis, G. *Two Instability Results in General Relativity*. Unpublished Princeton University doctoral dissertation (2018).

Mumford, D. "Ruminations on Cosmology and Time." *Notices of the American Mathematical Society* 68 (2021) 1715–1725.

Muthukrishnan, S. "Unpacking Black Hole Complemenarity." arXiv:2211.15650

Natterer, F. *The Mathematics of Computerized Tomography*. Society for Industrial and Applied Mathematics (2001).

Newman, R. "The Global Structure of Simple Space-Times." *Communications in Mathematical Physics* 123 (1989) 17–52.

Norton, J. "Observationally Indistinguishable Spacetimes: A Challenge for Any Inductivist." In G. Morgan (ed.), *Philosophy of Science Matters: The Philosophy of Peter Achinstein*. Oxford University Press (2011) 164–176.

Nussbaumer, H. and L. Bieri. *Discovering the Expanding Universe*. Cambridge University Press (2009).

Olver, P. *Applications of Lie Groups to Differential Equations*. Springer (1986).

Olver, P. *Introduction to Partial Differential Equations*. Springer (2014).

O'Neill, B. *Semi-Riemannian Geometry with Applications to Relativity*. Academic Press (1983).

Oxtoby, J. *Measure and Category: A Survey of the Analogy Between Topological and Measure Spaces* (second edition). Springer (1980).

Padmanabhan, T. "Statistical Mechanics of Gravitating Systems." *Physics Reports* 188 (1990) 285–362.

Page, D. "Is Our Universe Likely to Decay within 20 Billion Years?" *Physical Review D* 78 (2008) 063535.

Page, D. "Observational Selection Effects in Quantum Cosmology." In C. Glymour, W. Wang, and D. Westerstahl (eds.), *Logic, Methodology and Philosophy of Science: Proceedings of the Thirteenth International Congress*. College Publications (2009) 585–596.

Page, D. "Possibilities for Probabilities." *Journal of Cosmology and Astroparticle Physics* 2022 (2022) 023.

Page, D. "The Born Rule Fails in Cosmology." *Journal of Cosmology and Astroparticle Physics* 2009 (2009) 008.

Parikh, M. I. Savonije, and E. Verlinde. "Elliptic de Sitter Space: $dS/\mathbb{Z}_2$." *Physical Review D* 67 (2003) 064005.

Patrangenaru, V. "Lorentz Manifolds with the Three Largest Degrees of Symmetry." *Geometriae Dedicata* 102 (2003) 25–33.

Pauli, W. *Theory of Relativity*. Pergamon (1958).

Peacocke, C. "The Limits of Intelligibility." *The Philosophical Review* 97 (1988) 463–496.

Peebles, J. *Cosmology's Century: An Inside History of Our Modern Understanding of the Universe*. Princeton University Press (2020).

Peleska, J. "A Characterization for Isometries and Conformal Mappings of Pseudo-Riemannian Manifolds." *Aequationes Mathematicae* 27 (1984) 20–31.

Penington, G. "Entanglement Wedge Reconstruction and the Information Paradox." *Journal of High Energy Physics* 2020 (2020) 002.

Penrose, R. "Conformal Treatment of Infinity." *General Relativity and Gravitation* 43 (2011) 901–922. Originally published in B. deWitt and C. deWitt (eds.), *Relativité, Groupes et Topologie. Lectures, Les Houches, 1963 Summer School of Theoretical Physics*. Gordon and Breach (1964) 564–584.

Penrose, R. "Cosmological Mass with Positive Λ." *General Relativity and Gravitation* 43 (2011) 3355–3366.

Penrose, R. "Some Unsolved Problems in Classical General Relativity." In S.-T. Yau (ed.), *Seminar on Differential Geometry*. Princeton University Press (1982) 631–668.

Penrose, R. "Time-Asymmetry and Quantum Gravity." In C. Isham, R. Penrose, and D. Sciama (eds.), *Quantum Gravity 2: A Second Oxford Symposium*. Oxford University Press (1981).

Penrose, R. "Zero Rest Mass Fields Including Gravitation." *Proceedings of the Royal Society of London: Series A*. 284 (1965) 159–203.

Petersen, P. *Riemannian Geometry* (third edition). Springer (2016).

Poincaré, H. "Sur les hypothèses fondamentales de la géométrie." *Bulletin de la Société Mathématique de France* 15 (1887) 203–216.

Poisson, E. and G. Will. *Gravity: Newtonian, Post-Newtonian, Relativistic*. Cambridge University Press (2014).

Railton, P. "Alienation, Consequentialism, and the Demands of Morality." *Philosophy and Public Affairs* 13 (1984) 134–171.

Railton, P. "Scientific Objectivity and the Aims of Belief." In P. Engel (ed.), *Believing and Accepting*. Kluwer (2000) 179–208. This is a revised version of P. Railton, "Truth, Reason, and the Regulation of Belief." *Philosophical Issues* 5 (1994) 71–93.

Ratcliffe, J. *Foundations of Hyperbolic Manifolds* (third edition). Springer (2019).

Realdi, M. "Relativistic Models and the Expanding Universe." In H. Kragh and M. Longair (eds.), *The Oxford Handbook of the History of Modern Cosmology*. Oxford University Press (2019) 76–119.

Reiris, M. "Stationary Solutions and Asymptotic Flatness. I." *Classical and Quantum Gravity* 31 (2014) 155012.

Rendall, A. *Partial Differential Equations in General Relativity*. Oxford University Press (2008).

Rendall, A. "The Einstein-Vlasov System." In P. Chruściel and H. Friedrich (eds.), *The Einstein Equations and the Large Scale Behavior of Gravitational Fields: 50 Years of the Cauchy Problem in General Relativity*. Birkhäuser (2004) 231–250.

Rendall, A. "The Nature of Spacetime Singularities." In A. Ashtekar (ed.), *100 Years of Relativity. Spacetime Structure: Einstein and Beyond*. World Scientific (2005) 76–92.

Renn, J. and J. Stachel. "Hilbert's Foundations of Physics: From a Theory of Everything to a Constituent of General Relativity." In J. Renn (ed.), *The Genesis of General Relativity* Volume 4. Springer (2007) 857–974.

Rescher, N. "Leibniz and Issues of Eternal Recurrence." In N. Rescher, *On Leibniz* (expanded edition). University of Pittsburgh Press (2013) 117–135.

Rindler, W. "Elliptic Kruskal-Schwarzschild Space." *Physical Review Letters* 15 (1965) 1001–1002.

Ringström, H. "Future Stability of the Einstein-Non-Linear Scalar Field System." *Inventiones Mathematicae* 173 (2008) 123–208.

Ringström, H. "Instability of Spatially Homogeneous Solutions in the Class of $\mathbb{T}^2$-Symmetric Solutions to Einstein's Vacuum Equations." *Communications in Mathematical Physics* 334 (2015) 1299–1375.

Ringström, H. "On Proving Future Stability of Cosmological Solutions with Accelerated Expansion." In L. Bieri and S.-T. Yau (eds.), *One Hundred Years of General Relativity: A Jubilee Volume on General Relativity and Mathematics*. International Press (2015) 249–266.

Ringström, H. "On the Future Stability of Cosmological Solutions to Einstein's Equations with Accelerated Expansion." In S. Y. Jang, Y. R. Kim, D.-W. Lee and I. Yie (eds.), *Proceedings of the International Congress of Mathematicians: Seoul 2014* Volume II. Kyung Moon SA (2014) 983–999.

Ringström, H. *On the Topology and Future Stability of the Universe.* Oxford University Press (2013).

Ringström, H. *The Cauchy Problem in General Relativity.* European Mathematical Society (2009).

Röhle, S. "Mathematische Probleme in der Einstein-de Sitter Kontroverse." Max-Planck-Institut für Wissenschaftsgeschichte Preprint 210 (2002).

Rovelli, C. *The Order of Time.* Penguin (2018).

Ruelle, D. "Ergodic Theory." In E. Cohen and W. Thirring (eds.), *The Boltzmann Equation: Theory and Applications.* Springer (1973) 609–617.

Ruetsche, L. *Interpreting Quantum Theories.* Oxford University Press (2011).

Ruetsche, L. *The Physics of Ignorance.* Forthcoming from Oxford University Press.

Sachs, R. and H-H. Wu. *General Relativity for Mathematicians.* Springer (1977).

Sánchez, M. "On the Geometry of Static Spacetimes." *Nonlinear Analysis* 63 (2005) e455–e463.

Sánchez, N. "Quantum Field Theory and the 'Elliptic Interpretation' of de Sitter Spacetime." *Nuclear Physics B* 294 (1987) 1111–1137.

Santander, M., L. Nieto, and N. Cordero. "A Curvature-Based Derivation of the Schwarzschild Metric." *American Journal of Physics* 65 (1997) 1200–1209.

Scannell, K. "Flat Conformal Structures and the Classification of de Sitter Manifolds." *Communications in Analysis and Geometry* 7 (1999) 325–345.

Schemmel, M. "The Continuity Between Classical and Relativistic Cosmology in the Work of Karl Schwarzschild." In J. Renn (ed.), *The Genesis of General Relativity* Volume 3. Springer (2007) 157–181.

Schiffrin, J. and R. Wald. "Measure and Probability in Cosmology." *Physical Review D* 86 (2012) 023521.

Schrödinger, E. *Expanding Universes.* Cambridge University Press (1956).

Schrödinger, E. "Irreversibility." *Proceedings of the Royal Irish Academy* 53 (1950) 189–195.

Schrödinger, E. *Science and Humanism: Physics in Our Time.* Cambridge University Press (1951).

Schulman, L. "Note on the Quantum Recurrence Theorem." *Physical Review A* 18 (1978) 2379–2380.

Schwarzschild, K. "On the Permissible Curvature of Space." *Classical and Quantum Gravity* 15 (1998) 2539–2544. Originally published as "Über das zulässige Krümmungsmaass des Raumes." *Vierteljahrschrift der Astronomischen Gesellschaft* 35 (1900) 337–347.

Scott, P. "The Geometries of 3-Manifolds." *Bulletin of the London Mathematical Society* 15 (1983) 401–487.

Shalev-Shwartz, S. and S. Ben-David. *Understanding Machine Learning.* Cambridge University Press (2014).

Sider, T. *Four-Dimensionalism: An Ontology of Persistence and Time.* Oxford University Press (2001).

Silberstein, L. "General Relativity without the Equivalence Hypothesis." *Philosophical Magazine* (1918) 94–128.

Simányi, N. "Ergodicity of Hard Spheres in a Box." *Ergodic Theory and Dynamical Systems* 19 (1999) 741–766.

Simányi, N. "Further Developments of Sinai's Ideas: The Boltzmann-Sinai Hypothesis." In H. Holden and R. Piene (eds.), *The Abel Prize 2013–2017.* Springer (2019) 287–298.

Sklar, L. *Space, Time, and Space-Time.* University of California Press (1974).

Skow, B. *Objective Becoming.* Oxford University Press (2015).

Smeenk, C. "Einstein's Role in the Creation of Relativistic Cosmology." In M. Janssen and C. Lehner (eds.), *The Cambridge Companion to Einstein.* Cambridge University Press (2014) 228–269.

Smeenk, C. "Inflation and the Origins of Structure." In D. Rowe, T. Sauer, and S. Walter (eds.), *Beyond Einstein: Perspectives on Geometry, Gravitation, and Cosmology in the Twentieth Century.* Birkhäuser (2017) 205–241.

Smeenk, C. "Testing Inflation." In K. Chamcham, J. Silk, J. Barrow, and S. Saunders (eds.), *The Philosophy of Cosmology.* Cambridge University Press (2018) 206–227.

Smeenk, C. "Trouble with Hubble: Status of the Big Bang Models." *Philosophy of Science* 89 (2022) 1265–1274.

Smoller, J. *Shock Waves and Reaction-Diffusion Equations* (second edition) Springer (1994).

Smoller, J., B. Temple, and Z. Vogler. "An Alternative Proposal for the Anomalous Acceleration." In L. Bieri and S.-T. Yau (eds.), *One Hundred Years of General Relativity: A Jubilee Volume on General Relativity and Mathematics.* International Press (2015) 267–276.

Sogge, C. *Lectures on Non-Linear Wave Equations* (second edition). International Press (2008).

Srednicki, M. and J. Hartle. "Science in a Very Large Universe." *Physical Review D* 81 (2010) 123524.

Srednicki, M. and J. Hartle. "The Xerographic Distribution: Scientific Reasoning in a Large Universe." *Journal of Physics: Conference Series* 462 (2013) 012050.

Stachel, J. "Lanczos's Early Contributions to Relativity and His Relationship with Einstein." In J. Brown, M. Chu, D. Ellison, and R. Plemmons (eds.), *Proceedings of the Cornelius Lanczos International Centenary Conference.* Society of Industrial and Applied Mathematics (1994) 201–221.

Stanford, K. *Exceeding Our Grasp: Science, History, and the Problem of Unconceived Alternatives.* Oxford University Press (2006).

Stephani, H., D. Kramer, M. MacCallum, C. Hoenselaers, and E. Herlt. *Exact Solutions of Einstein's Field Equations* (second edition). Cambridge University Press (2003).

Sternberg, S. "Review of: *Imagery in Scientific Thought* by Arthur I Miller." *Mathematical Intelligencer* 8/2 (1986) 65–74.

Straumann, N. *General Relativity* (second edition). Springer (2013).

Strominger, A. *Lectures on the Infrared Structure of Gravity and Gauge Theory.* Princeton University Press (2018).

Susskind, L. and L. Thorlacius. "Gedanken Experiments Involving Black Holes." *Physical Review D* 49 (1994) 966–974.

Susskind, L., L. Thorlacius, and J. Uglum. "The Stretched Horizon and Black Hole Complementarity." *Physical Review D* 48 (1993) 3743–3761.

Szász, D. "Boltzmann's Ergodic Hypothesis: A Conjecture for Centuries?" In D. Szász (ed.), *Hard Ball Systems and the Lorentz Gas.* Springer (2000) 421–448.

Tao, T. *Poincaré's Legacies* Part I. American Mathematical Society (2009).

't Hooft, G. "Black Hole Unitarity and Antipodal Entanglement." *Foundations of Physics* 46 (2016) 1185–1198.

't Hooft, G. "Dimensional Reduction in Quantum Gravity." In A. Ali, J. Ellis, and S. Randjbar-Daemi (eds.), *Salamfestschrift.* World Scientific (1994) 284–296.

Thompson, M. *Life and Action: Elementary Structures of Practice and Practical Thought.* Harvard University Press (2008).

Thorne, K. and R. Blandford. *Modern Classical Physics: Optics, Fluids, Plasmas, Elasticity, Relativity, and Statistical Physics.* Princeton University Press (2017).

Thurston, W. *Three-Dimensional Geometry and Topology.* Princeton University Press (1997).

Tipler, F. "General Relativity and the Eternal Return." In F. Tipler (ed.), *Essays in General Relativity: A Festschrift for Abraham Taub.* Academic Press (1980) 21–37.

Tits, J. "Le principe d'inertie en relativité générale." *Bulletin de la Société Mathématique de Belgique* XXXI (1979) 171–197.

Tooley, M. "A Defense of Absolute Simultaneity." In W. L. Craig and Q. Smith (eds.), *Einstein, Relativity and Absolute Simultaneity.* Routledge (2008) 229–243.

Tumulka, R. "Long-Time Asymptotics of a Bohmian Scalar Quantum Field in de Sitter Space-Time." *General Relativity and Gravitation* 48 (2016) 2.

Unruh, W. and R. Wald. "Information Loss." *Reports on Progress in Physics* 80 (2017) 092002.

Valiente Kroon, J. *Conformal Methods in General Relativity.* Cambridge University Press (2016).

van Dongen, J. and S. de Haro. "On Black Hole Complementarity." *Studies in History and Philosophy of Modern Physics* 35 (2004) 509–525.

van Fraassen, B. *An Introduction to the Philosophy of Time and Space* (second edition). Columbia University Press (1985).

Van Raamsdonk, M. "Lectures on Gravity and Entanglement." In J. Polchinski, P. Vieira, and O. DeWolfe (eds.) *New Frontiers in Fields and Strings.* World Scientific (2017) 297–351.

von Neumann, J. "Proof of the Ergodic Theorem and the H-Theorem in Quantum Mechanics." *The European Physical Journal H* 35 (2010) 201–237. Originally published as "Beweis des Ergodensatzes und des H-Theorems in der neuen Mechanik." *Zeitschrift für Physik* 57 (1929) 30–70.

Wald, R. "Asymptotic Behaviour of Homogeneous Cosmological Models in the Presence of a Positive Cosmological Constant." *Physical Review D* 28 (1983) 2118–2120.

Wald, R. *General Relativity*. University of Chicago Press (1984).

Wald, R. *Quantum Field Theory on Curved Spacetimes and Black Hole Thermodynamics*. University of Chicago Press (1994).

Wald, R. " 'Weak' Cosmic Censorship." In A. Fine, M. Forbes, and L. Wessels (eds.), *PSA: Proceedings of the Biennial Meeting of the Philosophy of Science Association* Volume 2. Philosophy of Science Association (1992) 181–190.

Wallace, D. "A Bayesian Analysis of Self-Undermining Arguments in Physics." Forthcoming in *Analysis*.

Wallace, D. "Recurrence Theorems: A Unified Account." *Journal of Mathematical Physics* 56 (2015) 022105.

Wallace, D. "Why Black Hole Information Loss Is Paradoxical." In N. Huggett, K. Matsubara, and C. Wüthrich (eds.), *Beyond Spacetime: The Foundations of Quantum Gravity*. Cambridge University Press (2020) 209–236.

Walter, S. "Mathematical Milky Way Models from Kelvin and Kapteyn to Poincaré, Jeans and Einstein." *Oberwolfbach Reports* 12 (2015) 2081–2082.

Weinberg, S. *Gravitation and Cosmology: Principles and Applications of the General Theory of Relativity*. Wiley (1972).

Weinstein, S. "Undermind." *Synthese* 106 (1996) 241–251.

Weyl, H. "On the General Theory of Relativity." *General Relativity and Gravitation* 41 (2009) 1661–1666. Originally published as "Zur allgemeinen Relativitätstheorie." *Physikalische Zeitschrift* 24 (1923) 230–232.

Weyl, H. *Raum-Zeit-Materie. Vorlesungen über allgemeine Relativitätstheorie*. Julius Springer (1918). Subsequent German editions in 1919, 1919, 1921, and 1923.

Weyl, H. *Space-Time-Matter*. Methuen & Co (1922). Translation of the fourth German edition of *Raum-Zeit-Materie* of 1921.

Weyl, H. "Über die statischen kugelsymmetrischen Lösungen von Einsteins ‚kosmologischen' Gravitationsgleichungen." *Physikalische Zeitschrift* XX (1919) 31–34.

Williamson, T. *Knowledge and Its Limits*. Oxford University Press (2000).

Winitzki, S. *Eternal Inflation*. World Scientific (2019).

Wolf, J. *Spaces of Constant Curvature* (sixth edition). Chelsea (2011).

Woodward, J. "Sensitive and Insensitive Causation." *The Philosophical Review* 115 (2006) 1–50.

Wooters, W. and W. Zurek. "A Single Quantum Cannot be Cloned." *Nature* 299 (1982): 802–803.

Wu, J. "Stability, Genericity, and 'Physicalness' in Spacetimes." Unpublished (2020).

Zimmerman, D. "Presentism and the Space-Time Manifold." In C. Callender (ed.), *Oxford Handbook of the Philosophy of Time*. Oxford University Press (2011) 163–245.

Index